The Amazing Story of Life

From the Big Bang To the Future of Humanity

TOM KENNEDY

Kennedy Science Productions, LLC
Albuquerque, NM

Printed in United States of America

First Printing, 2019
ISBN-10: 1-945331-05-4
ISBN-13: 978-1-945331-05-3

Kennedy Science Productions, LLC
www.ksciproductions.com

FOR MY WIFE

Jen,

My best friend and life-long companion,
I could never have done this without you.
You're super awesome!
I love you so much!
Thanks for making sure I enjoy every moment of life.
Especially with pizza and beer.

CONTENTS

ACKNOWLEDGMENTS

I could have never done this on my own. Most importantly, I thank my family including my wife for her unwavering support, my parents, and my brother and sister-in-Law have all been instrumental in supporting me at every step in life. John Rogers for reviewing the first draft and providing valuable feed-back. My friends, Ian Latella, Eric Schaad, Mason Ryan, Jay Fuller, Steve Poe, Blair Wolf, Michael Collins, Bart Kicklighter, Hudson Cheng, and others of which I have had numerous adventures and late-night discussions with that helped me grow as a person and a scientist. And lastly, my PhD advisor Tom Turner who has been both a mentor and a friend.

Introduction

Why would I start the story of life with the Big Bang, an event that took place almost 10 billion years before the origins of life on our planet? Perhaps, it's because I always wanted to be an astronomer. In the early 1980s, I read the most influential book in my life, *Cosmos* by Carl Sagan. As I read through the pages, it gave me a sense of wonder about the universe, my first notions of science, and a sense of our place in the universe.

I remember Sagan calling us "star dust" because the carbon, oxygen, and nitrogen in our bodies were all forged in the heart of stars that lived and died billions of years ago. Even the building blocks to those elements have a more ancient origin going back to the Big Bang and the origins of the universe. These fundamental building blocks of matter, protons, neutrons, and electrons along with all the energy that will ever exist were created in an instant of time 13.7 billion years ago. Although, you as a person, may be only a few decades old, those tiny building blocks forming every atom in you and the rest of the universe, are 13.7 billion years old.

When I look into the night sky, it gives me a sense of wonder about our connections to the universe. Sagan Is right, we are star dust, quite literally. Our solar system formed about 4.6 billion years ago, a second or third generation star system rich with heavier elements like carbon, nitrogen, and oxygen the main ingredients of life. These elements were created by nuclear fusion inside long dead stars and released to the galaxy at the end of their lives. However, the creation of these the heavier elements was just the beginning of life's amazing story.

We don't know exactly when, where, or how life began. Although, once it got a foothold, life has been going strong ever since. Based on our current evidence, the origins of life occurred between 3.5 and 3.8 billion years ago, but it could have been earlier, perhaps 4 billion years ago. We may not know the exact time when life began, but life began simply, and we are descendants of those first living cells. Go back far enough, and all life is related by sharing a common ancestry. Remarkably, we can read the history of our species and our relationship to all other life forms written in our DNA.

A few years ago, two friends once again changed my views of life. It happened during happy hour when a friend of mine, who happened to be a geologist told me, "life is an extension of geology". Like Sagan's "we are star dust" claim, he changed the way I thought about life, a light bulb went off as I realized life's connection to the Earth. If life is an extension of geology, then life had to come from some geological process. About the same time, another friend of mine, also at the same happy hour, suggested I read Nick Lane's *Life Ascending*. I went home, ordered the book on Kindle and three days later, I've never thought of life in quite the same way.

Like most people, I thought about the question "what is life?", or what characteristics separate life from something non-living, like a rock. In my opinion, that's not the best way to understand life. Instead of asking the question, "what is life" or "what are the characteristics of life", perhaps, a better way to think about life begins by asking "what does life do?" The question, "What is life?" implies that life is a noun, or a list of characteristics that you might memorize in a freshman biology course. For example, I can remember finishing my PhD in biology at the University of New Mexico when someone asked me a simple question, "what is life?" At first, I attempted to recite the list of five characters taught in our book. I came up with life reproduces, metabolizes, moves, and that was about it. After getting through two or three of these characteristics described in our intro text book, I couldn't remember the others.

I felt embarrassed that with a PhD in biology, I could not define life based on text book knowledge. For the next few years, I glossed over the lesson of what is life in my classes, feeling unsure about the definitions in the book. But at the time, I didn't have a better idea. It wasn't until our happy hour conversations and reading *Life Ascending* and the other books by Nick Lane, I began to realize life was something more than a list of characteristics; life is an action, it does something. Life is a verb, not a noun; so, life is an action, then it is those actions that makes life different from something non-living.

Armed with a better sense of what life does, its connections to the universe and the Earth, I wanted to write my own science book about life. The goal is to describe life amazing story, even its humble beginnings on the ancient Earth. Throughout this book, think about life's connections to the world around it and what life must do to be alive, regardless of how small, large, or bizarre it appears. To fully appreciate life, we need to know a little astronomy, physics, chemistry, geology, and climatology, basically a little of all the sciences.

Life's' connections to the universe began long ago. It may be difficult to believe, but astronomical events starting with the Big Bang, a single moment in time when all the energy and matter in the universe were created set stage for life . Later, exploding stars billions of years ago created the heavier elements that would form our Earth and the building blocks of life. Today, we still depend on our sun for a constant supply of energy to fuel life.

Life is connected to the Earth. In an ancient ocean sometime around 3.7 to 4 billion years ago, the building blocks of life assembled into complex molecules until life emerged as an extension of geological processes. The first living organisms used energy from the environment to create order from chaos making us fundamentally different from anything non-living. Life remains out of equilibrium with its environment, where death returns us to equilibrium. Contrary to new age beliefs, you never want to be in equilibrium with your surroundings because that means you would be dead.

Since its beginnings, life has evolved to become a major force shaping the Earth's surface. We can thank our oxygen-rich atmosphere to tiny bacteria that first evolved photosynthesis billions of years ago, making oxygen as a byproduct. These early photosynthetic organisms paved the way for the evolution of complex life including plants and animals that would come to dominate the Earth's surface.

Reproduction makes the evolution of life possible because reproduction passes information from one generation to the next, ensuring the continuity of life. In each generation, small errors called mutations occur when genetic information isn't perfectly copied. These errors provide the variation among organisms vital for the evolutionary forces of natural selection to act upon. Since its beginnings, life has evolved as genes have been passed down for countless generations, slowly changing over time. In fact, you are the result of an unbroken lineage going back about 3.8 billion years to the first living organisms.

No organism is an island, existing alone or isolated. All life must interact with its environment to acquire nutrients and energy to create order. Energy and molecules flow through our bodies like water flows through a river, connecting every living organism to the Earth. Unlike energy which cannot be recycled, the elements forming molecules are continuously recycled, made available by special types of living organisms. Every carbon atom in you was once in a molecule of carbon dioxide in the atmosphere until it was fixed by a plant into a sugar by harnessing a practically unlimited source of energy from the sun.

Like all other animals, we are connected to the Earth, although most remain unaware of those connections and how our activities harm the planet. Humans are unlike any species that has come before us. Our population has rapidly grown due to cheap energy and the technology to exploit it. Unfortunately, we are causing rapid changes from habitat loss and degradation to climate change. As a result, the 6th mass extinction, rivaling the end of the Mesozoic when the dinosaurs disappeared, has begun.

What's our fate? Will humans survive the 6th mass extinction, or suffer the fate of the dinosaurs? Only time will tell. Regardless of our actions or inactions, the legacy of humans will be written into the geological record and the future evolution of life on this planet.

The Nature and Limitations of Science

Finally, after thousands of years, we finally have modern science to understand the world. At its root, science means the desire to know. In modern times, science has evolved into a process that allows us to understand the natural world. Science begins by making observations and asking questions. Unfortunately, sometime between elementary and high school, most people stop observing the natural world around them, let alone ask questions about it. By the time most of us are adults, we simply take the world for granted.

I live in Albuquerque, NM where we have a 10,500-foot mountain east of town, the Rio Grande River flowing through the middle of town, and recently extinct volcanoes west of town. After teaching thousands of students, few have ever asked the simplest questions about the landscape in our town: what caused the Sandia Mountains to form, how old are they, why are there fossils of ancient marine life on top of the mountain, or why are there volcanoes west of town.

Mankind has always been curious about the natural world, even if most individuals are not. How many myths and creation stories have been dreamed up to answer the most basic questions about where we come from, or to understand our place in the universe. For the first time, in the history of our species, we have science, a method that allows us to understand the world as it is, not through the lens of ideology or fanciful beliefs. Additionally, the efforts of thousands of scientists have accumulated scientific knowledge over time through a repeating and self-correcting process. Hypotheses, which are proposed explanations for a set of observations, get tested, and either kept or discarded based on the data. Even the most cherished hypothesis can be quickly proven false and discarded with a single observational fact.

Contrary to popular belief, science has its limits, it can only explain the natural world. When I use the term, "natural world", it has a specific meaning with major implications. Things that we can potentially observe, measure, and test for define the natural world. Using these criteria places limits on science, which is a good thing. If not, we'd have to entertain any idea to explain our world, no matter how implausible or absurd it is. If there is no conceivable way to make an observation or test for a phenomenon, then it's outside the realm of science and may even be pseudoscience. Pseudoscience makes claims about the world that "sound or appear scientific" but does not actually follow the conventions of science. Examples of pseudoscience include intelligent design, or aliens built the ancient world.

Sounding like science, pseudoscience leads the world astray, but it doesn't work simply because it can't be tested. Imagine a scenario where someone said the Earth rests on the back of a giant, invisible sea turtle swimming through the cosmos. You can't test for the turtle's existence, and you can't disprove or prove it's there. To put it bluntly, those kinds of fanciful ideas are beyond the scope of science because you can't observe, test for, and you can't potentially disprove it. It's a good thing we limit the scope of science, that way we don't get preoccupied trying to disprove an unlimited stream of untestable ideas.

I'm totally fine with that, I don't want to use my time thinking about things I won't have a chance at understanding or ever figuring out. I would rather use my time thinking about things that I have a shot at understanding like the origins of life, or the effects of climate change on the southwest. It gives me comfort that scientists never stop observing, asking questions, and testing their hypothesis to learn about our world, no matter how difficult the question. Because of the continuous efforts of scientists following the conventions of science, our knowledge and understanding of the universe grows removing the need for supernatural explanations to explain things we don't understand.

In addition to pseudoscience, other misconceptions regarding science creep into popular beliefs. Much of this stems from how we teach science in school: as a collection of facts to be memorized, or hypotheses are merely educated guesses. Scientific studies do accumulate facts, but more importantly, science is a process and a way of thinking about the natural world. Hypotheses are borne out of our observations and prior knowledge, they are predictions or proposed explanations, not guesses.

Science is not another belief system based on unverifiable stories lacking in support. Evidence that we can verify through observations or experimentation forms the bases of science. As we improve our understanding of the natural world, science makes additional predictions based on our hypotheses, theories and laws. If they fail to accurately predict outcomes, they get modified or scrapped. Throughout this book, you will notice that scientific theories, including the Big Bang theory, origins of life, evolution by natural selection, or global climate change, make certain predictions that have been repeatedly verified based on scientific evidence.

Science has shed light on the life's story. In this book, I share some of the amazing things life does as it has evolved and diversified over the eons. Throughout this book, think about our connections to the universe and the Earth. I begin with the Big Bang to show how astronomy and then geology set the stage for life to emerge. Once it began, life's journey from single-celled bacteria to human civilization has been long but filled with time periods where bacteria dominated the oceans for a billion years, to the rapid diversification of animals, shortly followed by the colonization of land and the greening of the Earth. Along the way life survived through planet-wide ice ages lasting millions of years, mass extinctions, volcanism, a few meteor impacts, and now it must survive humans.

Chapter 1

Origins of the Universe
and the
Earth

It is far better to grasp the universe as it really is than to
persist in delusion,
However satisfying and reassuring.

Carl Sagan

Introduction

When I'm hiking on the west side of the Sandia Mountains near Albuquerque, NM, I can't help but think that the mountain range is young for mountains, about 10 million years old. However, the granite forming the mountain formed over 1.5 billion years ago, making it older than the dinosaurs, even older than the first animals. To further put this into perspective, the granite forming the Sandia Mountains was formed when all life existed merely as single-celled organisms. If you were to go back in time to the formation of the Sandia granite you would not recognize the Earth, other than it was mostly covered in water.

In many ways, we share similarities to the granite forming the mountains. Although, no one would mistake a rock as living, yet many the same elements found in life also forms rocks. Those elements are ancient, older than the Earth itself. They were formed inside stars that exploded billions of years ago spewing their insides to the universe. Even the elements are made of more fundamental building blocks, just three basic subatomic particles called protons, neutrons, and electrons.

For the first time in the history of our species, we live in a time when we have the technology to understand the nature of the universe and explain our origins without invoking super-natural explanations. To accompany our technological advances allowing us to observe tiny cells or distant galaxies, we have also developed a methodology, commonly known as the scientific method, providing a way for us to understand the natural world. Importantly, we should all remember that science is not good or bad, conservative or liberal, it's a process of understanding the natural world. It's also important to know that contrary to popular beliefs, science is limited in its scope to understanding the natural world.

The Universe Began with a Bang!

How many times has the question, where did we come from, been asked? Over the course of humanity, thousands of myths, stories, and religions popped up to answer this fundamental question about our own existence. Some of the oldest stories of our civilization center on creation myths. Now for the first time in the history of our species we can use science to answer this age-old question.

Before the 1920s, most scientist assumed the universe was ageless, it had been around forever and was mostly unchanging. Even Einstein attempted to modify his General Theory of Relativity to accommodate the view of an unchanging, and eternal universe. Years later, he called it the biggest blunder of his life. Ironically, before admitting to his blunder, it was his General Theory of Relativity that led a Belgian Catholic Priest and astronomer named Georges Lemaitre to predict an expanding universe and the Big Bang in 1927, two years before Edwin Hubble.

Published in Belgian, Lemaitre's paper on the expanding universe went unnoticed. However, support for an expanding universe began to appear in the 1920s when a new generation of large telescopes began making detailed observations of distant galaxies. Perhaps the most well-known astronomer of the time was Edwin Hubble, the same astronomer that the Hubble Telescope is named after.

Prior to these large telescopes, we didn't know much about the universe, including its age, size, or the nature of galaxies. At the time, most thought the Milky Way included much of the universe, and without much information, there really wasn't any reason to believe it had a beginning or an end. The first glimpses of the vast size and age of the universe emerged when Hubble began making a series of observations on faint "fuzzy-like" objects known as galaxies. After making many observations over several years, it became clear that galaxies were large clusters of stars beyond our own galaxy.

Hubble made an important observation about galaxies, he noticed that the light coming from galaxies was shifted to the red end of the light spectrum, we call it a redshift. It works like this; light moves in waves and the wavelength determines the color of light. Longer wavelengths are reddish and shorter wavelengths are blueish. Hubble also showed that the redshift of galaxies was not random, it increased with distance. The further a galaxy was from us, the more its light was red-shifted.

If all the light from galaxies is red-shifted, this means they are moving away from us. It's called the Doppler Effect, you may not be familiar with the name, but you experience it almost daily. Imagine standing on a roadside watching a car approach, the sound is high pitched because the sound waves are compressed. Once the car passes you, the sound is lower pitched because the sound waves are elongated. Because light travels in waves, it also experiences the Doppler Effect. Therefore, light from an object moving away from us would appear reddish because its light waves were being stretched.

Based on this simple observation, astronomers realized that almost every single galaxy is moving away from us because their light waves are red-shifted. If all the galaxies are moving away from us, that means we live in an expanding universe. But, the other important implication happens if you go back in time, the universe would be contracting. Go back far enough, there would be a time when the universe didn't exist, thus placing a finite age on the universe.

Since Lemaitre and Hubble's findings in the late 1920s, a steady stream of evidence continued to support an expanding universe with a beginning and finite age. However, the idea that the universe began with a cataclysmic explosion was so ground breaking that the British astronomer Sir Fred Hoyle referred to the hypothesis as "the Big Bang" in an attempt to ridicule the theory. In an ironic twist of fate, the name stuck. Now we call it the Big Bang Theory, or simply the Big Bang. By the 1990s, astronomers calculated the universe's age to be approximately 13.8 billion years old.

Like other scientific theories, the evidence strongly supports the Big Bang theory, no data has yet to disprove it, and it remains our best explanation for the origins of the universe. The use of the term 'theory' in science has specific meaning. A scientific theory is a broad, powerful explanation for a set of observations. In this case, the Big Bang Theory explains the observations of an expanding universe and predicts a finite age of the universe. Most good theories, including the Big Bang, also make additional testable predictions. If the universe began with the Big Bang, then there should be additional evidence to support this theory. For example, if the universe began with a cataclysmic explosion, then shouldn't some of that residual energy still be present as cosmic background radiation?

The discovery of this cosmic background radiation was accidentally made by scientist from the Bell Laboratories in the 1960s, when a state-of-the art radio telescope detected "background-noise" from all parts of the sky. If you've ever heard static on the radio, that's similar to the cosmic background radiation. At the time, the scientists had no idea what was causing the "hum" from the radio telescope. They checked all the wiring, removed bird and rodent nests, and even had to rule out the implausible idea that the former Soviet Union was beaming radio waves at us. After several frustrating years and numerous attempts to "fix" the radio telescope, it became clear that the noise was the residual energy of the Big Bang. They discovered the cosmic background radiation predicted by the Big Bang theory, thus lending additional support to the theory.

Hubble's observation of redshifts from distant galaxies is a scientific observation, which can be taken as a fact. A fact is something that exists or actually happens, much like it's a fact the sun sets in the west and rises in the east. To a scientist, observational facts lead to more questions, such as why are the galaxies moving away from us. To answer their questions, scientists make hypotheses, which are proposed explanations for their observations. To be a scientific hypothesis, it should be potentially testable and falsifiable. I use the word potentially because it may not be possible to test some hypothesis because it costs too much money, we don't yet have the technology, or it may not be ethical.

Originally, the expanding universe started as a hypothesis proposed by George Lemaitre in 1927 based on General Relativity. Like other hypothesis, it was testable, potentially falsifiable, and better yet, it made testable predictions about the universe that were verified by observations, including the cosmic background radiation. For nearly 100 years, scientific observations and additional tests continue to support the Big Bang theory predicting an expanding universe with a finite age. The accumulation of evidence helped the Big Bang theory to become the most well-supported scientific theory to explain the origins of the universe. Since its start, it has yet to be disproved by a single observation.

The development of the Big Bang theory also illustrates the process of science: observations lead to questions that are turned into testable hypotheses. With further study, hypotheses become well supported, grow in their scope, and eventually become accepted as theories. Making observations, formulating hypotheses, testing hypotheses, and using data to refine our explanations defines the scientific method.

Although science remains a powerful tool for understanding the natural world, it does have limitations. Some questions, such as, "what existed before the Big Bang" may be difficult to address simply due to a lack of data because the information has been lost, or may be very difficult to obtain. Despite these obstacles, scientists continue to answer difficult questions by putting forth and testing hypotheses. We discard the ones that don't work and keep the ones that are well supported. Just because a question is difficult, does not mean it is not worth asking. If you never ask the difficult questions, they won't ever get answered. It is the tenacity of scientists to keep asking questions, no matter how hard, that keeps pushing the boundaries of knowledge.

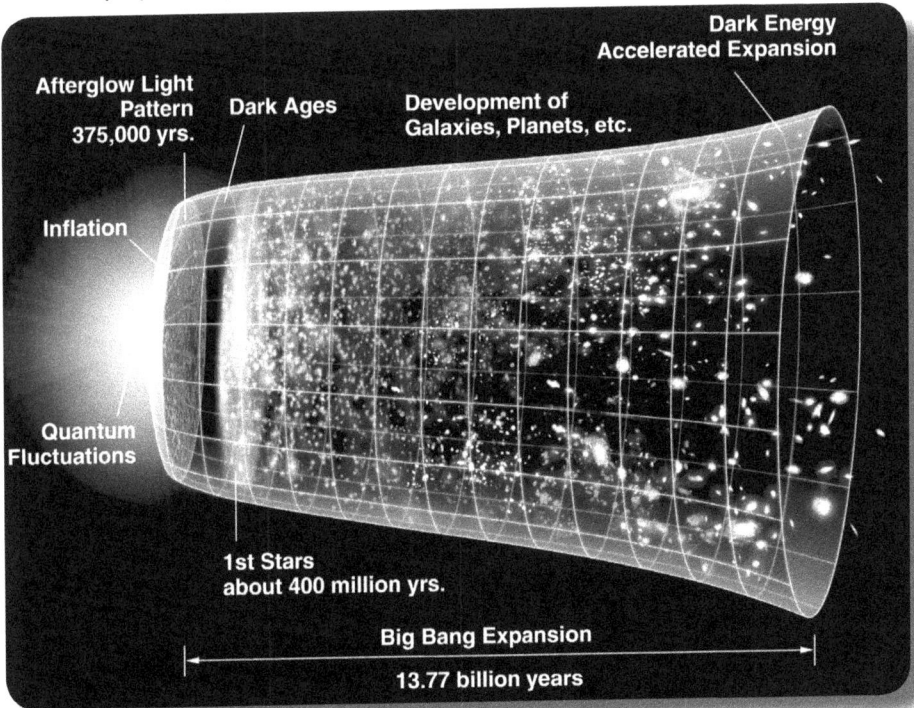

Artistic illustration showing the expansion of the universe beginning with the Big Bang, followed by rapid expansion known as inflation

What is Matter and Energy?

Imagine that in a single instant, all the matter and energy in the universe was created. This leads us to a fundamental question about the nature of our universe, what exactly is matter and energy? To begin, we define matter as something that occupies space and has mass. This is known as the Pauli Exclusion Principle which states: No two particles of matter can occupy the same place at the same time. There are many particles of matter, but the ones most important to our understanding of biology are three subatomic particles known as protons, neutrons, and electrons that combine to form atoms.

Atoms are the smallest particles of matter that have the properties of an element, such as oxygen, gold, or calcium. The atomic structure of an atom consists of the nucleus with positively charged protons and neutrally charged neutrons, both roughly equal in mass. Each element comes with an atomic number and an atomic mass. The number of protons in the nucleus determines the atomic number of an element. Change the number of protons and you have a different element. The number of protons and neutrons determine the atomic mass of elements. Negatively charged electrons with a mass one thousandth of a proton exist in a cloud surrounding the nucleus. In any atom, the number of electrons always equals the number of protons. For example, hydrogen, the simplest element, which accounts for 75% of the atoms in the known universe, is comprised of a single proton and one electron.

The Big Bang also created all the energy in the universe. Energy is a property of objects that allows them to effect change on the universe. Similar to life, energy is best described by what it does rather than what it is. Simply, we define energy as the ability to do work. Broadly speaking, there are two types of energy, potential and kinetic energy. Kinetic energy is the energy of motion, the faster an object is moving the more energy it has. Imagine, you're walking down the hallway texting on your cell phone and accidentally bump into a wall, it's unlikely to cause much harm. But, if you were running really fast into a wall, the outcome could be grim because you have more kinetic energy.

Potential energy is the stored energy of an object, which can depend on its relative position to other objects. For example, drop a glass from a few feet above the floor, it will likely break due to its relative position to the floor. In contrast, when you tip over your glass on a table it probably won't break. There are many ways to store energy, including, water behind a dam, stretched rubber bands, batteries, or chemical energy stored in bonds that form molecules.

You can't break the Laws of Thermodynamics

Scientific law or scientific theory; what's the difference? If you're like me, I've been confused more than once trying to wrap my ahead around the differences between facts, laws, and theories, and it took me some time to understand the difference. However, before I continue, not everyone agrees with my definition, but I think they work most of the time. It's important to know the differences because we use theories and laws to make sense of our world, and we accumulate facts along the way.

To understand the difference between laws and theories, I'll use gravity as an example. Isaac Newton discovered the law of gravity in the 1680s when he determined that an object will fall to the Earth at a specific rate each time it is dropped. Newton's law of gravity makes a prediction regarding the rate an object will fall to the Earth. However, Newton's law of gravity does not explain how gravity works, or what gravity is. It took another two centuries and an intellectual leap for the nature of gravity to be explained. In 1915, Einstein explained how gravity works in his General Theory of Relativity. He predicted that the mass of large objects causes a curvature in space-time. Since its publication over 100 years ago, the General Theory of Relativity has been repeatedly verified through observations and experimental tests and remains our best explanation for how gravity works. General Relativity was revolutionary and years ahead of its time, it predicted an expanding universe and black holes before there was any observational evidence to support them.

In physics, the laws of thermodynamics govern how energy behaves. Without exception, all life is subject to the laws of thermodynamics. In total, four laws of thermodynamics explain energy, but the first two laws are the most important for biology. The first law of thermodynamics tells us that one form of energy can be converted into another form.

For example, when you start a fire, potential energy in the wood gets converted to kinetic energy that we see as light and feel as heat. The first law of thermodynamics also tells us that energy cannot be created nor destroyed, instead it remains constant! This is awesome, because the first law would imply that we should never run out of energy.

Unfortunately, there is the second law of thermodynamics, which states that every time we use energy, the total entropy of the universe increases. Entropy is a measure of disorder or randomness. If you were to take your clothes out of the dryer and throw them on the bed, then that would be high entropy, or disorder. If you were to hang your clothes in a closet and organize them by color, then you work to lower entropy by creating order.

Unfortunately, the universe will grind to a state of maximum entropy, slowly decaying until no usable energy remains. The universe will end with a whimper when the last star winks out existence, but the universe will continue to expand forever in infinite darkness. Sounds bleak, but none of us will be around to witness the end.

Getting back to entropy and the second law of thermodynamics, we know if you quit cleaning your house, it would reach a state of equilibrium where everything was just randomly distributed in each room. It requires energy to do work to maintain a clean and organized house. However, by organizing your house, you directly add to the total entropy of the universe.

Remember, every time energy is used, entropy increases, no exceptions, it's an unbreakable universal law. Because of this annoying law of nature, energy cannot be recycled and perpetual motion machines cannot exist. If it weren't for the second law of thermodynamics and increasing entropy, you could eat a single meal in your life time and recycle the energy forever. But, that's not how our universe works, life in must have a continual supply of energy to exist. Cut life off from energy, and it will die as it reaches equilibrium, a topic I'll discuss further in Chapter 3.

The Origin of the Elements

Hanging on the walls of many science classrooms is the Periodic Table of Elements. The chart organizes about 92 naturally occurring elements arranged in order of their atomic number. Hydrogen and helium, the simplest and most abundant elements in the universe formed shortly after the Big Bang, sit at the of the chart. Where did the other 90 elements come from? The answer to their origins lies in stellar processes that we can observe today. Based on our best evidence, it took about 200 million years after the Big Bang for the first stars to form leading to the creation of the other elements.

Stars form when clouds of hydrogen collapse from the force of gravity causing their cores to heat up under the immense pressure. Once the cores reach about 14 million degrees Celsius the protons gain so much kinetic energy, they overcome their natural repulsion and stick together forming the element helium. Nuclear fusion occurs anytime protons or protons and neutrons become forced together growing the size of an atom's nucleus. In addition to forming new elements, nuclear fusion makes neutrons, which together with protons forms the nucleus of elements.

Nuclear fusion releases an enormous amount of energy, enough to counteract the force of gravity. Once fusion begins, the star stabilizes in size as it stops collapsing. The light emitted from the sun results from nuclear fusion. Once nuclear fusion begins, a star like our sun, is born. At the center of our sun, the temperature is about 15 million degrees Celsius and 250 billion times the pressure at sea level!

Stars don't last forever, eventually they run out of hydrogen and when they do, nuclear fusion stops. At this point, the star will begin to collapse from the force of its own gravity, further heating its core causing helium atoms to fuse into heavier elements. Through a series of repeated contractions of stars at the end of their lives, additional elements such as carbon, nitrogen, oxygen, sodium, chlorine, calcium, and potassium are created over millions of years, or tens of thousands of years for very large stars. Eventually, some of these massive stars will begin to produce Iron-56. Once this happens, the energy released from nuclear fusion is no longer sufficient to prevent the rapid collapse of the star from its own gravity. The collapse of these large stars is so rapid, nearly 70,000 km/s, that the outer layers hit the core and then rebound causing an enormous explosion called a supernova.

Supernova explosions are among the most impressive astronomical events in the universe, in a few moments they will release more energy than our sun will release in millions of years! When stars explode, the elements formed from nuclear fusion get scattered into space. Interestingly, the force of the explosion contains enough energy to make elements beyond iron including gold, silver, mercury, lead, and uranium. Just think, the gold or silver in our jewelery was created by fusion during an enormous stellar explosion! Life on Earth, including us, is 96% hydrogen, carbon, nitrogen, and oxygen. With the exception of hydrogen, all the elements in our bodies were created in stars and spread by supernova explosions that took place more than 5 billion years ago. We are stardust, the remains of ancient stars that ended their lives in spectacular explosions billions of years ago.

Origin and Age of the Earth

The origin of our solar system began approximately 5 billion years ago when a large star reached the end of its life and exploded as a supernova scattering its contents of heavier elements into local space. At the time our solar system was a nebulae, nothing but a cloud of gas and dust. A shock wave may have triggered the nebula to collapse forming the sun, the planets, and all the other comets, asteroids, and dwarf planets. Our solar system is a second or third generation solar system containing many more elements created from the remnants of earlier supernova explosions. During its formation, the sun picked up most of the matter in our solar system. Smaller fragments containing rocks and other metals also began to grow and accrete into smaller bodies eventually forming the planets.

Although the sun accounts for 99% of the mass of the solar system, 8 planets eventually formed along with thousands smaller bodies, such as comets, asteroids, and dwarf planets with Pluto being the most famous one. The inner four planets are small and rocky, three of which have atmospheres. The four outer planets are much larger gas giants because they lack a rocky surface like the Earth. In recent years, astronomers have found some evidence that another large planet may have formed along with the eight planets we see today, but was ejected out of the solar system 4 billion years ago.

Radioisotope dating places the age of the Earth and the solar system to be approximately 4.6 billion years old. To understand radioisotope dating, we need a little knowledge regarding atomic structure. Recall that the protons, neutrons, and electrons are the building blocks of elements. If the number of protons changes, you have a new element. However, element vary in the number of neutrons and this does not change its chemical properties. Different numbers of neutrons form an isotope, but the atomic number remains the same. However, the atomic mass of an element changes with the number of neutrons.

Some configurations of protons and neutrons are very stable, forming stable isotopes. For example, carbon atoms always contain 6 protons, and the vast majority of carbon atoms contain 6 neutrons for an atomic mass of 12. Rarely, a carbon atom will contain 7 neutrons with an atomic mass of 13. Both Carbon-12 and Carbon-13 are stable isotopes because they don't decay into other elements. In fact, the stable isotopes, such as carbon-12 are so stable we don't know how long they could last. Estimates place the lifespan of atoms somewhere between 10^{25} to 10^{34} years, an unimaginably long time considering the universe is only about 13.7×10^9 years old. To put that into perspective, an atom could last 10 trillion times longer than the current age of the universe.

Some isotopes have too many or too few neutrons making them unstable. They are known as radioisotopes because they emit subatomic particles at a specific rate as they change into other elements. Luckily for us, the decay rates of radioisotopes does not change over time. Additionally, each radioisotope has a unique half life; for example, the half-life of Carbon-14 is about 5,730 years. Here's how it works; suppose you have a pound of a radioactive substance like potassium-40. It has a half-life of 1.25 billion years, so after 1.25 billion years you would have 0.5 pounds, after 2.5 billion years, you would have 0.25 pounds. Every time 1.25 billion years pass by, you would have half the amount you started with.

Every element has radioactive isotopes that decay at specific rates, and the half-lives of different isotopes range from a few minutes to billions of years as in the case of Uranium-238 with a half-life of approximately 4.5 billion years. We can determine the age of the solar system or the Earth by using radioactive isotopes found in meteorites, ancient rocks on Earth, and moon rocks returned by the Apollo Missions. So far, all the different isotopes support a similar age for the Earth and solar system at approximately 4.6 billion years.

How do we know that radioactive isotopes have always decayed at the same rate? This is a good question and took years to answer. Based on over one hundred years of observations and experiments conducted by thousands of scientists, we know that radioactive decay is based on the statistical decay rate of a population of radioisotopes and doesn't change over time. Additionally, the conditions under which most radioisotopes formed don't exist on the Earth, so there are no forces present that would alter their decay rate.

Another reason why we have good reason to believe that the half-lives of radioisotopes do not change is based on the principle of uniformitarianism. Uniformitarianism makes the assumption that the same natural laws that apply on Earth are the same throughout the universe, at any point in the past, or at any point in the future. In the case of radioactive decay, this means that the rate Uranium-238 decays into lead, with a half-life of 4.5 billion years, is the same today, the same in the past, and the same anywhere else in the universe.

Uniformitarianism is a concept borrowed from geologists in the early 1800s wishing to understand how current geological processes can inform us about the past. Over time, the assumption has been repeatedly supported by observations and experimental verification. To date, there are no scientific reasons, observations, or experiments that have refuted uniformitarianism, this basic scientific principle has withstood the test of time and scientific verification. To state it bluntly, we don't have any reason not to believe it. Uniformitarianism is quite useful to understanding the universe because it would be almost impossible to understand the natural world if the rules were constantly changing.

4.6 Billion Years Ago to Present
A *Ridiculously* Brief Summary of Earth's History

The Earth is ancient, not as old as the universe, but still far older than most of us can easily comprehend. To organize Earth's history, geologists have created geological time units based on major changes in geological processes or based on major events in the history of life. The largest unit of time is the eon, which can last for billions of years. Eons are subdivided into eras that span tens of millions of years, eras are subdivided into periods, which are further subdivided into epochs. Below, I provide a brief description for some of these time periods.

Most of us have a difficult time comprehending extremely large numbers such as the age of the universe or the age of the Earth. To help us understand just how old the Earth is, let's take an imaginary road trip back in time. Imagine that one millimeter, the width of the period at the end of this sentence, represents a year in time. Therefore, someone who has lived 25 years would represent 25 millimeters in length, or 2.5 centimeters, which is about an inch. Not a very long road trip! The average life span in the US is about 85 years, which would be 85 millimeters or about 3.3 inches. To go back a thousand years, you would move one meter, or a little over three feet. To go back to the late Pleistocene Epoch 22,000 years ago when glaciers dominated the northern hemisphere, we would only have to travel 22 meters, or about 72 ft.

To travel back in time to the origins of the first modern humans 200,000 years ago, we would walk a mere 200 meters, a little longer than two football fields. If we were to travel back in time to witness the extinction of dinosaurs 65.5 million years ago, we would have to travel 65.5 kilometers or about 39.3 miles. If we were to travel back in time to observe the origins of the first animals sometime between 550 and 600 million years ago, we would have to travel about 550-600 km (330-360 miles) a decent 6-hour road trip on an interstate highway. To go back to the beginning of the Earth we would have to travel 4600 km (2700 miles), a trip this long could take you from Albuquerque, NM to New Brunswick Canada and take almost 40 hours of driving time! That's an enormous distance considering one millimeter represents an entire year!

The first three eons of Earth's history are determined mostly by astronomical and geological processes, such as the end of the heavy bombardment of asteroids and comets at the end of the Hadean Eon. However, beginning with the Phanerozoic Eon 542 million years ago, the geological time scales are also determined by life in addition to geological changes to the Earth, a testament to the importance of life on the planet. Very little evidence remains of the first eon, but as we get closer to modern times, we have a more complete picture of the geology, climate, and life's diversity. We can use clues found in ancient rocks to piece together a picture of the Earth's history. Below is a brief summary of the ages of the Earth, beginning with the first eon and ending with our current epoch.

Hadean Eon (4.6-4 Billion years ago): Named after the Greek god of the Underworld, the Earth in its infancy was a hellish place. Its surface was most likely largely molten and continually bombarded by large meteors. Within less than a hundred million years of its formation, the Earth was hit by another planet-like body nearly the size of Mars vaporizing a large part of the planet that later formed the moon. The surface was almost certainly liquefied and any atmosphere was blown away into space. By the end of the Hadean, the surface had cooled enough to be covered in water and an atmosphere rich in carbon dioxide was present. It's unknown whether or not life was present in the Hadean, mostly because there are no rocks or minerals present from this time period due to tectonic activity and the constant erosion and weathering of the surface.

Archean Eon (4.0-2.5 Billion years ago): Named after ancient Greek for beginning, or origin, the Archean spanned nearly 1.5 billion years of Earth's history. Solid evidence for life exists from colonies of single-celled bacteria called stromatolites, dating back to about 3.7 billion years ago. Other indirect evidence indicates that life may have begun around 3.8-4 billion years ago. Throughout the book, I will use 3.7 billion years as the origin of life, based on the direct evidence. It's important to point out that not everyone agrees with the indirect evidence for the first signs of life. If you were to visit the beginning of the Archean, you would not fare so well because the atmosphere lacked free-oxygen (O_2). However, evidence from ancient rocks indicates that cyanobacteria evolved early in the history of life and began to pump oxygen into the atmosphere as a byproduct of photosynthesis. Over time, these tiny bacteria changed the composition of the atmosphere as free-oxygen became much more abundant. It is generally accepted that volcanism was much more prevalent, bringing lighter rocks to the surface. By the end of the Archean, the first continents existed, but their size and distribution remains largely unknown.

Proterozoic Eon (2500 – 542 Million years ago): Named after ancient Greek for "earlier life", the Proterozoic was the longest eon spanning almost 2 billion years, or 43% of the Earth's history. During this time, atmospheric oxygen levels steadily rose due to the relentless photosynthetic activity of cyanobacteria. Fossil evidence indicates that the first eukaryotic cells and multicellular life evolved during this time, their evolution almost certainly made possible by the rise in free-oxygen levels in the atmosphere. It was during this eon that the Earth began to take on a recognizable appearance.

Phanerozoic Eon (542-present): Named after ancient Greek for "visible life", the Phanerozoic Eon began with the Cambrian Explosion named after the rapid appearance and diversification of animal life about 542 million years ago. Since the Cambrian Explosion, life has evolved to fill every corner of the Earth. Currently, about 2 million species have been described, and some researches have estimated the total diversity to be over 8 million species. Even more amazing, today's diversity represents only about 1% of all the diversity that has existed since the beginning of life. The Phanerozoic Eon is divided into three Eras, the Paleozoic, Mesozoic, and Cenozoic.

Paleozoic Era (542-251 MYA): Meaning "old life", the Paleozoic Era began with the rapid appearance of multicellular life, or specifically the first fossils of trilobites. By 500 million years ago, most modern groups of animals including arthropods, mollusks, and chordates were present. By the end of the Paleozoic, life had greatly diversified and conquered the land for the first time in the Earth's history. Plants had evolved the ability to grow tall, forming vast forest and creating new opportunities for diversification of animals. Fish had evolved and diversified to dominate the seas. Also, one small lineage of fish made the transition to land as they evolved into tetrapods. By the end of the Paleozoic, the early ancestors of reptiles and mammals were present.

The Paleozoic Era lasted for nearly 300 million years, abruptly ending 251.9 million years ago with the largest mass extinction of the Phanerozoic Eon called the End-Permian. Massive volcanism may have been responsible for the mass extinction. A region known as the Siberian Traps cover an estimated 2 million square kilometers of volcanic rock, which would cover nearly 25% the size of the lower 48 states in rock over a kilometer thick. The massive volcanic activity may have altered the climate and chemistry of the oceans driving nearly 85-95% of all species to extinction in a few hundred thousand years, the blink of an eye in geological time.

Mesozoic Era (251-65.5 MYA): Meaning "middle life", the Mesozoic Era followed on the heels of the End-Permian extinction. It took nearly 20 million years for the diversity of life to recover. Despite starting slowly, the Mesozoic Era is best known for its reptiles. The most famous were the dinosaurs that roamed the Earth for nearly 170 million years. The Mesozoic Era also witnessed the evolution of flowering plants, modern mammals, and birds, all of which survived the mass extinction event ending the Mesozoic Era. Evidence suggests that the End-Mesozoic extinction may have been caused by a one-two-punch starting with massive volcanism altering the climate and the oceans, and ending with the death blow from a large meteor hitting the Earth 65.5 million years ago, sealing the fate of the dinosaurs along with more than half of the world's diversity.

Cenozoic Era (65.5 MYA – present): Meaning "new life", the Cenozoic saw the continued diversification of flowering plants, insects, fish, birds, and mammals. Some call it the age of mammals, but there are actually more species of birds than mammals, and more fish than birds and mammals combined! During the Cenozoic Era, the Earth's climate has slowly cooled resulting in ice ages with large ice caps covering the poles. The Cenozoic Era can be divided into three periods, the Paleogene, Neogene, and Quaternary.

Paleogene Period (65.5 – 23 MYA): The Paleogene Period followed the aftermath of the dinosaur extinction. Diversity recovered quickly as birds and mammals quickly diversified and competed for terrestrial dominance. Mammals eventually edged out birds to become the top predators and diversified into a great many forms including flying bats, aquatic whales, and terrestrial predators.

Neogene Period (23 – 2.6 MYA): Modern groups of mammals appeared during the Neogene including the ancestors to modern humans. North America became connected to South America when the Isthmus of Panama formed. This connection allowed for a Great American Interchange of life, causing the extinction of many South American species that had been isolated from the rest of the world for tens of millions of years. The isthmus also altered the ocean currents, causing the Earth to further cool and intensifying the current ice age. Not everyone agrees on the start date of the current ice age, some place it starting nearly 34 million years ago when Antarctica became covered in ice sheets.

Quaternary Period (2.6 MYA – Present): The defining events of the Quaternary Period include the growth and retreat of large ice sheets as the Earth oscillated between cold periods of maximum glaciation and relatively quiescent interglacials of reduced ice-cover that we enjoy today. Modern humans also evolved during the Quaternary Period, which is divided into two epochs, and a third that is gaining in popularity.

Pleistocene Epoch (2.588 MYA – 11.7 thousand years ago): Defined by the extensive glaciation of the Quaternary period. It ended with the end of the last major glaciation. Modern humans evolved in Eastern Africa and spread throughout the world.

Holocene Epoch (11.7 thousand years ago – present): Meaning entirely recent, it began at the end of the last major glaciation and encompasses the entirety of written records and human civilization. The Holocene has been a time of relative climate stability.

Homogenozoic Era (Present - ?) I made this one up, it's based on the fact the world is currently experiencing a 6th mass extinction and the diversity after this modern extinction event will be homogenized due to the spread of invasive species. I make the case for a new era at the end of the book.

Anthropocene Epoch (present - ?): Never before has there been a single species with the same impact humans unleashed on the Earth in such a short time. The rapid burning of fossil fuels, releasing carbon into the atmosphere that had been buried for hundreds of millions of years, is warming the Earth at unprecedented rates. Combined with the degradation of ecosystems resulting from overexploitation, the spread of invasive species, and rapid climate change, we are causing an extinction event that rivals the end of the Mesozoic era when the dinosaurs were wiped off the Earth. Because of our impact on the Earth, there has been a recent movement among some geologists and biologists to name the current time the Anthropocene Epoch due to the impacts of man's activities on the Earth. At the end of the book, I return to this concept, and suggest that we may actually be entering a new geological era, which I would like to call the Homogenozoic Era.

Throughout the book, I will use these time periods as a reference when I discuss major evolutionary innovations, including the evolution of life, photosynthesis, aerobic respiration, eukaryotes, and the rise and diversification of plants and animals.

Eon	Era	Period	Epoch	m.y.
Phanerozoic	Cenozoic	Quaternary	Holocene	
			Pleistocene	2.6
		Neogene	Pliocene	
			Miocene	23
		Paleogene	Oligocene	
			Eocene	
			Paleocene	66
	Mesozoic	Cretaceous		
		Jurassic		
		Triassic		251
	Paleozoic	Permian		
		Carboniferous		
		Devonian		
		Silurian		
		Ordovician		
		Cambrian		542
Precambrian		Proterozoic		2500
		Archean		3800
		Hadean		4600

Chapter 2

The Origins of Life

Life began 3.8 billion years ago, necessarily about as simple as it could be, because life arose spontaneously from the organic compounds in the primeval oceans.

Stephen J. Gould

Introduction

What led to the origin of life some 3.8 billion years ago? This may be one of the most interesting and challenging questions biologists have been asking for decades. For years, theories explaining the origins of life have continually been proposed, tested, discarded, or modified due to our increasing knowledge of the conditions of the early Earth. Developing new theories, modifying them, even discarding a popular theory reflects the process of science by reminding us of science's dynamic nature. Science is a repetitive and self-correcting process, ultimately leading to a more complete and accurate understanding of the natural world.

To understand how life got started, I find it helpful to ask the question, "What does life do?" Thinking about life as a process is different than describing life as a set of characteristics. Perhaps, a better way to describe life is by what it does rather than what it is. Think about it, life does something, and knowing what it does helps us create better theories of life's origins.

At the smallest scale, life emerges from complex chemical reactions using energy to create order. To appreciate the role of chemistry in life, I include some basic chemistry, a hard task to sell as something interesting, but I believe a little knowledge of chemistry can go a long way to fully appreciating how life works. In the second part of this chapter, I review the history of theories proposed to explain life's origin beginning with Darwin in the 1860s. I place them in a historical context to show how scientific theories can change to incorporate new findings.

By now, you may be asking: how do we know our current theories won't be thrown out with some new discovery? Is everything we currently know just going to be changed in the future? These are valid questions, and sometimes our best theories get overturned almost overnight with a single new observation. But, as science progresses, fewer theories will be overturned with a growing body of knowledge generated by science. Eventually, through our continuous efforts, science eliminates hypotheses and theories that don't work leaving us with a better understanding the universe.

Would we recognize alien life if we saw it?

Perhaps, the most iconic scene in movie history occurs in *Star Wars* when a young Luke Skywalker enters the seedy underworld of the Mos Eisley Spaceport Cantina. Walking into the cantina introduced thrust him into a much larger universe full of strange aliens. The vastness of our universe will likely prevent us from finding intelligent aliens in our life time. However, we are sending probes to other planets and moons in our solar system and using powerful satellites to search for exoplanets outside the solar system with the hope of discovering alien life.

To me, finding extraterrestrial life, or life that did not originate on the Earth, would be one of the most significant events of our lifetime. Finding alien life anywhere in the universe would be humbling, yet a triumphant achievement for humanity to know for certain that we are not alone in the universe. The search for extraterrestrial life begs the question, would we recognize alien life if we saw it? To me, the answer is yes, especially if we know how to recognize when something is living. To do this, we should ask the question, what does life do, rather than what is life. If we know what life does, then no matter how alien looking, we would recognize it as life.

So, what does life do? What makes something living? At its most basic level, life uses energy create order. The constant flow of energy breathes life into the universe. Life must have energy; without it, life could not exist. In fact, if you cut the energy supply off to an organism it will die. Therefore, death, like life, is a process where entropy takes over. Slowly, organic molecules decay and eventually the remains of the organism will reach equilibrium with its environment. Its elements return to the geological cycles of the Earth, perhaps to be used again by another animal.

Contrary to popular belief, you never want to be in equilibrium with your environment, because you would no longer be living. Life is an open system requiring a constant flow of energy, at least most of the time. For example, we take in potential energy stored in the foods we eat. Eventually, most of the energy will leave our bodies as heat.

No living organism is an island, all life, from the smallest cell to the largest animal, must acquire the energy from its surroundings to create order. Ecology, studies how life interacts with its environment to survive and reproduce, which I will discuss in Chapters 11 and 12. I'll start with the basics here. Autotrophs (auto = self, troph = feeding) form the foundation of all ecosystems. They make energy and nutrients available to everything else by using energy to fix carbon dioxide into organic molecules, life sugars. On land, plants are the most common autotrophs. They use the energy in sunlight to make sugars from water and carbon dioxide. Autotrophs are vital to life because they transform the energy in sunlight and store it in organic molecules, making energy and nutrients available to other organisms in an ecosystem.

In contrast to autotrophs, heterotrophs get energy from organic molecules, such as the sugars in a piece of fruit, or by eating another animal. Heterotrophs are quite varied and include many bacteria, all animals, and fungi. Recall that energy can't be recycled because every time it is used, it becomes degraded and lost as heat, a form of energy not well suited for doing work. Therefore, energy must constantly flow through life in an ecosystem starting with the autotrophs.

Unlike energy, the elements in an ecosystem, including carbon and nitrogen, are continuously recycled. Decomposers, including many types of fungi, break down organic matter and recycle it, making it available to plants, who once again use the sun's energy to make the elements available to everything else.

However, it requires more than an input of energy for life to exist. A frozen pizza has every element and molecule required for life. Yet, you can heat a frozen pizza in the microwave all you want and life won't emerge from your efforts to warm dinner. The input of energy is just as likely to break down molecules as it is to build new ones.

Using energy to create life requires a myriad of chemical reactions. Inside every cell, millions of coordinated chemical reactions occur every second. They break down large molecules into smaller ones and use the smaller building blocks to make larger molecules. These reactions also harness the energy released from the break-down of large molecules to do work. You can think of work as using energy to move something. Imagine cleaning your room, by picking up your clothes you are expending energy to do work. However, by creating order inside your room, you actually increase the total disorder of the universe. Doing work, requires energy, but most of the energy gets converted to heat and lost to the environment. That's why you sweat while working, your body needs to remove the excess heat generated by all the energy being used.

The loss of energy as heat whenever we do work satisfies that pesky second law of thermodynamics; every time we spend energy for work, entropy increases. Although life creates order, but it does so at the level of the organism, and can only do so as long as there is a constant supply of energy coming in. So don't worry, life doesn't violate any laws of nature.

Our metabolism includes millions of chemical reactions taking place inside of cells every second of every day. These reactions are not random, but are highly controlled, breaking down molecules, extracting energy, and using it to make whatever the cell needs. Controlling our metabolism and storing the information to make the enzymes that speed up all those chemical reactions requires a special molecule called nucleic acids. Cells use two types, DNA and RNA. DNA stores the information required to make all the molecules required for our metabolism. Its cousin, RNA carries out the directions of DNA, more on the roles of DNA and RNA in Chapter 5.

For there to be a continuity of life, there must be reproduction. Without reproduction, life would emerge only to perish shortly afterwards. Reproduction copies information stored in DNA, creating a new organism with that information. Copying the information stored in DNA inevitably leads to mistakes, which can be detrimental, neutral, or even beneficial. Through natural selection, beneficial mistakes create the diversity in life allowing it to evolve, continually adapting to a changing environment. In 1859 Charles Darwin published his book, *On the Origin of Species*, where he presented his lifetime's work supporting the theory of evolution by natural selection explaining how life changes over time to adapt to its environment.

Let's summarize what life does. From a physics point of view, life uses energy to create order taking it out of equilibrium with its environment. Or, life is an island of low entropy surrounded by a sea of chaos. From a chemistry point of view, life extracts energy from its environment through chemical reactions to make new, and more complex molecules. From a biological point of view, genetic information directing the activities of life is copied to make new organisms, thus ensuring the continuity of life.

From a Darwinian point of view, life evolves over time as it adapts to exploit new resources or adapt to a changing environment. At its heart, life is a process that interacts with its environment to extract energy and nutrients to create order.

Lastly, I'd like to return to the question I posed at the beginning of the chapter: would we be able to recognize alien life if we saw it? For me, the answer is yes. If we identify the processes of life on another moon or planet, then we almost certainly have life regardless of its chemistry or alien appearance.

If we know what life does, it helps us propose testable hypotheses and develop new theories to explain the conditions leading to the first living organisms and look for those conditions on other planets and moons. Since Darwin's time, theories to explain the emergence of life on Earth have been put forth. Before diving into the theories for the origin of life, I'd like to discuss the importance of water and carbon for life.

Life Requires Liquid Water

The chemistry of life includes countless chemical reactions where molecules are broken down and new molecules are formed. On this planet at least, all of life's chemical reactions depend on liquid water to provide a medium for the reactions to occur, simply because water is a good solvent. Without water, molecules would rarely encounter each other, and life would never arise. Contrary to popular belief, water is not a universal solvent. If water were a universal solvent, then it would dissolve anything it touched on contact, including my Diet Pepsi bottle.

Another important property of water for life is the pH, a measure of how acidic or alkaline the water is based on the concentration of hydrogen ions [H^+]. Ions are charged particles; they can be positively charged or negatively charged. Hydrogen is the simplest element with one proton and one electron. If you remove the electron, you have a positively charged ion, in this case specifically, a proton. The pH scale is a measure of the **p**otential of **h**ydrogen (pH) ions and ranges from 0 to 14 where 7 is neutral, a pH less than 7 is acidic and a pH greater than 7 is alkaline or basic.

Strongly acidic or alkaline water contains more chemically reactive ions. You can see the effects of acidity if you squeeze lemon juice on a piece of uncooked fish. After a few moments, you will notice that the texture of the fish will change as the lemon juice reacts with the proteins forming the muscles in the fish. The chemistry of our cells is finely tuned to an optimum pH where small deviations quickly create problems for the life

In addition to being a good solvent, water has several other important characteristics important for life. Water is cohesive, it sticks to itself. The ability to stick together makes water liquid at room temperature, or why water molecules stick together to form a drop of rain. Water also sticks to other objects, a property known as adhesion. Adhesion and cohesion are especially important for plants, it's part of the reason why water moves from their roots to their leaves, sometimes several hundred feet away in tall trees, without the plant spending any energy to actively move the water.

Water has a high heat capacity, meaning it takes a lot of energy to change its temperature. I become keenly aware of this property when I'm hungry and waiting for a pot of water to boil to make some hearty mac and cheese. The pot will quickly heat up, but the water takes much longer because it is absorbing lots of energy. In fact, water's high heat capacity makes it a greenhouse gas holding energy and moderating air temperatures. For example, cloudy nights don't get as cold as clear nights, that's because water holds energy from the day and moderates the temperature. If you've ever lived by the ocean, then you know the coast doesn't experience the large temperature changes of inland areas.

The Chemistry of Life is Based on Carbon

Every organic molecule is based on carbon. Similar to water, special features of carbon make it vital for life allowing it to form complex molecules that perform many functions including, speeding up chemical reactions, storing information, storing energy, communication, forming barriers, or used in movement, just to name a few examples. The versatility in carbon lies in its ability to form four stable chemical bonds with other elements, including the ability to repeatedly bond to itself, creating an almost endless number of molecules. In fact, that the entire field of organic chemistry is based on studying carbon compounds.

We don't know how many organic molecules exist in nature, some estimates place it in the millions. Despite the almost infinite possibilities, most organic molecules fall into one of four basic groups of macromolecules, carbohydrates, proteins, nucleic acids, and lipids (which include fats). Not every organic molecule easily fits into one of these four categories, including pigments and vitamins. For example, the pigment chlorophyll, giving leaves their green color, and captures the energy in sunlight during photosynthesis. In animals, several types of pigments called melanin is responsible for eye, hair, and skin color. Other organic molecules required for the proper functioning of cells are vitamins, including vitamin C and B.

Because life depends on breaking down and building organic molecules, I have included a brief description of the main functions of each macromolecule. Also, understanding how these molecules work can improve our health, help us create new medicines, prevent chronic health problems, or even slow down the aging process.

Lipids form a broad class of organic molecules with different functions. For example, fats store energy, especially in animals. Sterols, including estrogen and testosterone allowing plant and animal cells to communicate long distances. Phospholipids form cellular membranes, which surround every cell. These unique molecules organize into bilayers where one end attracts to water and another end that avoids water. The importance of cellular membranes cannot be over stated, they form the barrier between the living inside of the cell and the non-living outside world. Without cellular membranes, life would remain little more than metabolically active rocks, unable to exist on their own.

Best known as source of energy, carbohydrates serve many other purposes. Drinking a Mountain Dew loaded with sweet tasting sugars provides some of us with an extra boost in energy to power through afternoon slumps or late night studying. Surprisingly, all life uses glucose for energy; another piece of evidence supporting that all life shares a common ancestor. Linking together glucose molecules makes complex carbohydrates. The most common ones include cellulose and starch, both made by plants. Plants store energy in starch to be used at later times like in the spring when plants need to rapidly grow, or risk being out-competed. For example, potatoes, store lots of starch, which we use to make mash-potatoes, French fries, or hash-browns. In contrast to starch, cellulose forms the cell walls of plants providing some of their support, yet animals do not use it for energy. To put this into perspective, glucose makes cellulose, the most common organic molecule on the planet, yet animals lack the ability to break down cellulose into glucose. Rather than make their own enzymes to digest cellulose, herbivores rely on microbes to breakdown cellulose. Although, cellulose passes through our digestive, it provides improves our digestive health and maintains healthy gut microbes.

Nucleic acids, including DNA and RNA store our genetic information and regulate the expression of our genes. Although quite large, DNA only uses four building blocks called nucleotides (Adenine, Guanine, Cytosine, and Thymine) whose specific sequence codes for every living organism of the planet. Imagine if the English language only used four letters. RNA was once thought to be an intermediate, carrying out the instructions held in DNA to make proteins. However, new evidence suggests that RNA, a close relative of DNA, was extremely important in the origins of life because of its ability to store information and promote chemical reactions.

Proteins are the work horses of the cell. With potentially millions of proteins and more discovered every day, they carry out a variety of functions. The varied role of proteins is truly amazing, they can be used for structural support, to speed up chemical reactions, for cellular communication, and some identify and destroy pathogens. Despite the vast number of proteins and their varied purposes, only 20 different building blocks called amino acids are used to make proteins. Imagine if the English language only had 20 words!

What Conditions Led to the Origin of Life?

In 1859 Darwin speculated that life began in a small warm pond millions of years ago. In the 1960s Francis Crick proposed that life

THE ORIGINS OF LIFE

began on Mars or somewhere outside our solar system and was brought to Earth on a meteor, a comet, or even by ancient aliens; a hypothesis known as panspermia. Crick believed that the conditions leading to life were so rare, it was unlikely to have occurred here. However, this scenario is unlikely and I prefer the simplest answer, life began on Earth sometime about 3.7-4 billion years ago. The reason for believing life began on Earth rather than somewhere else, is because all the ingredients needed for life are right here; a constant supply of energy, liquid water, lots of carbon, and a geologically active planet.

What were the conditions that led to life's origins? Scientists have been studying and testing hypotheses to answer this question for decades. Perhaps the most challenging problem lies in reconstructing what the Earth was like at the dawn of life. Was the Earth covered in oceans? Were the oceans salty? Were they acidic or alkaline? Were continents present? What was the composition of the atmosphere? The Earth is a dynamic planet with its slow, but steady plate tectonics and constant erosion removing the evidence of the Earth's past, makes it hard to know what the Earth was like billions of years ago. As a result, competing theories describing the conditions of the early Earth have been proposed.

Despite competing theories describing the Earth's past, most agree that water covered the surface 3.8 billion years ago, much like today. We know this because the oldest rocks and minerals found were clearly formed in a watery environment. Additional evidence from these ancient rocks indicates that carbon dioxide, nitrogen, and water dominated the atmosphere 4 billion years ago with potentially with small amounts of methane, hydrogen, sulfur dioxide, and hydrogen sulfide. Unlike today, extensive volcanic activity releasing carbon dioxide made it a major part of the atmosphere. The extra carbon dioxide was a good thing because the sun in its early days was only about 75% of its current brightness. These high levels of atmospheric carbon dioxide kept the Earth warm, preventing a deep freeze delaying the emergence of life.

If we traveled back in time to the Hadean Eon four billion years ago, we wouldn't live long on the surface because Earth's ancient atmosphere lacked free-oxygen (O_2, which is a gas and I will refer to as oxygen). In fact, it's unlikely that any planetary atmosphere would have large amounts of oxygen unless something put it there. Although animals need oxygen to live, it is extremely reactive, easily breaking down complex organic molecules in to smaller, and more stable molecules such as carbon dioxide. Fortunately, the early atmosphere, lacking in oxygen, made it favorable to the formation of complex organic molecules required for life.

The early Earth was geologically active, warm enough for liquid oceans, and surrounded by a thick atmosphere devoid of oxygen and rich in greenhouse gases. Life emerged through chemical evolution and the direct consequence of geological processes. We also know that the molecules of life require carbon and the chemistry of life takes place in water. Without carbon or liquid water, it is unlikely that the processes of life could have emerged. Therefore, when searching for extraterrestrial life, we look for liquid water and carbon.

Prior to Darwin, people thought that life arose through spontaneous generation, a process where life simply popped out of non-living material. This was not an absurd hypothesis, but stemmed from observations that fly larvae, commonly known as maggots, seemed to "appear out of nowhere" on rotting plant or animal material. Louis Pasteur experimentally disproved this theory in the early 1860s.

It wasn't until 1953 when Stanley Miller and Harold Urey developed the prebiotic soup theory, providing the first testable hypothesis of abiogenesis, the evolution of life from non-living substances. In 1986, the Nobel Laureate Walter Gilbert proposed the RNA world theory predicting that abiogenesis began with self-replicating RNA molecules. In the 2000s, a new theory predicting that metabolism evolved before replication, has emerged and is quickly becoming widely accepted because it answers the problems with earlier theories of abiogenesis. The continual development and refinement of theories of abiogenesis spanning the last 150 years shows how our understanding of difficult problems changes as we gain scientific knowledge.

Miller-Urey and the Prebiotic Soup Theory

The atmosphere four billion years ago likely contained carbon dioxide, water, and nitrogen, but importantly, it lacked oxygen. To determine whether organic molecules could form under these primitive conditions, a graduate student named Stanly Miller conducted the famous Miller-

Urey experiments with his advisor Harold Urey in the early 1950s. In their experiment, they created a system that contained sea water and a mix of gases they thought represented the early atmosphere including water (H_2O), methane (CH_4), ammonia (NH_3), and hydrogen (H_2). To carry out the experiment, they heated the water and added energy in the form of electricity to the mix. After a week, a brown sludge began to appear in the catchment tube. They collected the substance and found the building blocks of life, amino acids along with other organic molecules!

Critics of the Miller-Urey experiments claimed they were a failure because they did not produce life. There is a big step going from the building blocks of life to an actual living organism. How could such an improbable event happen? Despite the criticisms of the Miller-Urey experiment, it was a success for several reasons. First, it was the first scientific tests for the origins of life, conclusively demonstrating that the building blocks of life are easily made from simple molecules. The Prebiotic Soup Theory that grew out of these early experiments was a valid scientific theory because it provided a potential explanation for chemical evolution leading to the origins of life. For over 30 years, the prebiotic soup theory remained the leading theory to explain the origins of life.

However, the prebiotic soup model does not explain how the organic molecules created in the "soup" could come together to form a cell. You can add carbon dioxide, nitrogen, and other small organic molecules to water, add some energy, and you will get the building blocks of life. However, adding energy doesn't mean you will keep getting more and more complex molecules, the energy that builds molecules just as easily break them down. This presents a problem, small organic molecules can be continuously made, only to be broken down again by the next lightening strike. Recall that a living organism uses energy to create order, so how would an ordered system, such as a cell, arise from a prebiotic soup of organic molecules?

The shortcomings of the prebiotic soup theory led researchers to keep asking questions, scrapping or modifying our hypotheses when they lacked support, it's the process of science. The Miller-Urey experiments of the 1950s, showed that the building blocks of life are easily formed, an important piece of information for the development of new hypotheses.

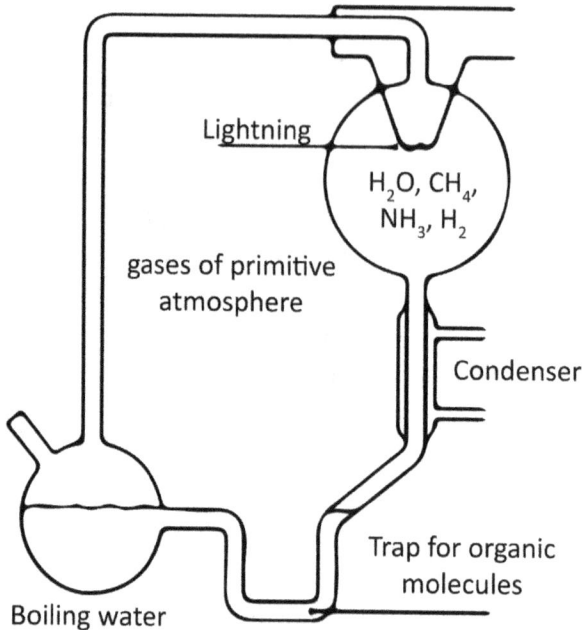

The Miller-Urey Experiment was similar to this set up where water was boiled in a flask, the water vapor then entered a gaseous chamber with electrical sparks to simulate lightning strikes, serving as a source of energy. The gaseous mix was then cooled by a condenser and the trap at the bottom collected the organic molecules.

The RNA World Theory

When it comes to life, biologists historically focused on reproduction and evolution. A key process of life, reproduction ensures its continuity and existence over time. When life reproduces, genetic information is replicated and forms the next generation. Based on the importance of reproduction and the findings of the prebiotic soup theory, the RNA World Theory explaining life's origin became widely popular in the 1980s when the Nobel Laureate Walter Gilbert published a commentary on recent findings in the prestigious scientific journal *Nature*. Although the theory's roots date back to the 1960s with the discovery of RNA, the unique properties of RNA forms the basis of the theory. RNA, a molecule similar to DNA, also stores genetic information. However, beginning in the late 1960s and through the 1980s, experiments showed that RNA causes certain chemical reactions similar to protein enzymes. What could be better to explain the origins of life? RNA, a molecule that stores information, catalyzes chemical reactions to make molecules, and perhaps most importantly, it catalyzes its own replication.

Additional evidence about the roles of RNA inside cells provided support for the RNA world theory. For example, RNA builds proteins, or specifically ribosomal RNA (rRNA). Because RNA carries genetic information and helps make proteins, the same proteins it carries information for, most thought that RNA preceded DNA as the first molecule to store information and replicate itself. The RNA world hypothesis places RNA and replication at the center of life rather than DNA. However, DNA being more stable than RNA is the reason why modern cells use DNA to store genetic information and it makes sense to store information in a more stable molecule.

From its beginning, the RNA world hypothesis quickly became a popular theory with wide-spread acceptance. It championed the idea that life began as a simple self-replicating molecule eventually evolving into the millions of life-forms we see today. By using results from the Miller-Urey experiments, scientists knew that the building blocks of life, including amino acids and nucleotides, formed naturally. The RNA world provided an explanation as to how life began, starting with self-replicating molecules in a prebiotic soup.

After a decade, the RNA world faced major challenges as several flaws in the theory came into question. First, how would enough RNA nucleotides form from the prebiotic soup to make larger molecules. Or, where did all the RNA come from? While RNA does self-replicate, the length of RNA depends on the concentration of its building blocks, RNA nucleotides. In the prebiotic soup model, the RNA nucleotides would be diluted, preventing RNA molecule from growing large, never reaching sizes required to make the simplest of proteins. It is also unclear how RNA got inside a membrane and began to direct chemical reactions required for metabolism.

Metabolism First Theory

In the late 1990s, scientists began to question the RNA World Theory as new evidence began to emerge. Experimental results showed that the length of RNA in a prebiotic soup would be too short to form complex proteins. Remember, that based on the principle of uniformitarianism, we know that chemical reactions would operate the same on the ancient Earth as they do today, the laws of the universe don't change over time. That's how we know that experimental results from today can be used to understand the past.

To put forth a new theory of abiogenesis, we needed new information and the fresh perspective from geologists and biochemists tackling the problem. In the early 2000s, major gains in our understanding of the chemical properties of the oceans and other geological processes 4 billion years ago led to new theories. Current evidence strongly suggests that the ancient oceans were acidic with a pH less than 7. Today, the oceans are alkaline with a pH averaging 8.1.

The metabolism first theory began with a remarkable discovery in 1977 when researchers using the deep sea submersible *Alvin* discovered hydrothermal vents in the Pacific ocean near the Galapagos Islands. Hydrothermal vents are found near volcanically active areas and spew out hot water along with chemicals and metals conducive to forming new molecules. Because the water from these vents is laden with iron sulfides, it looks like black smoke bellowing from chimneys, hence the name black smokers. Most surprisingly, these hydrothermal vents, found in the cold depths of perpetual darkness of the ocean, were teaming with life. This was a totally new type of ecosystem; instead of using energy from sunlight, it used the energy coming from these vents.

With the discovery of these unique ecosystems surrounding the hydrothermal vents, scientists began to ask whether life began in similar environments. Volcanism was much more common billions of years ago making these hot vents commonplace. The hydrothermal vents would have provided energy and a favorable environment for making organic molecules necessary for life to emerge from chemical evolution. The significance of the hydrothermal vents as a source of energy is important because the first living organisms harnessed energy from geological processes to make organic molecules from carbon dioxide and hydrogen gas and other small molecules.

The discovery of the hydrothermal hot vents provided a vitally important clue to life's origins because it solved problems with the prebiotic soup and RNA world theories. Hydrothermal vents provided a constant source of energy to make complex molecules and concentrate them into small areas, similar to modern cells. But, like other theories, metabolism first comes with its own problems. For one, the hydrothermal vents last only a few decades before becoming inactive. That may not be enough time for life to evolve. Secondly, they are hot, the water coming out them can reach up to 600 degrees Fahrenheit, hot enough to kill almost any living organism.

A second major discovery occurred in 2000 when an expedition of scientists, once again in *Alvin*, discovered a different type of hydrothermal vent called an alkaline vent. The most popular one, called "The Lost City", has about 40 chimneys, some nearly 200 feet high, and perhaps thousands of years old. Alkaline vents differ in several important ways from the black smokers discovered in the 1970s. Instead of being hot, they are warm, about 90-110 degrees Fahrenheit. Additionally, alkaline vents remain geologically active for thousands of years, potentially providing enough time for chemical evolution leading to abiogenesis.

In the last decade, William Martin and Michael Russel have suggested that life first evolved in similar alkaline vents. Perhaps the most important aspect of the alkaline vents lies in their geochemistry. Here's how these vents work; warm alkaline water (pH around 8-9) rich in hydrogen gas and dissolved minerals percolates through the chimneys. Naturally occurring tiny spaces fill the alkaline vents like holes in a sponge forming tiny chemical reactors. The alkaline-rich water would move through these small spaces depositing minerals that facilitate chemical reactions.

Let's make the connection to the importance of what is happening in these alkaline vents. Recall that a living organism uses metabolic processes to keep out of equilibrium with its environment, a condition that requires a constant supply of energy to maintain. The alkaline vents were out of equilibrium with the surrounding acidic oceans, providing a natural source of energy. Additionally, metals like iron and magnesium lined the inside acting as catalyst, speeding up chemical reactions. The tiny chambers inside the vents, about the size of a bacteria, concentrated the organic molecules allowing for chemical evolution to take place. This scenario would not violate the second law of thermodynamics because these tiny chambers were open systems with a constant supply of energy and materials. There's more to this story and I will return to it in Chapter 3.

The *Lost City* is an alkaline hydrothermal vent in the Atlantic Ocean and represents the type of geological formation where life may have begun.

The metabolism first theory predicts that geological processes caused chemical evolution to occur by creating a system out of equilibrium with the environment. From chemical evolution, life emerged, starting with metabolism. This means the first living organisms on were little more than a metabolically active rock! Importantly, this theory of metabolism first enjoys supported from experimental evidence in labs showing that chemical evolution would occur under these circumstances. To date, this is the best theory explaining the origin of life. The ability to use DNA to direct complex chemical reactions and reproduction may have evolved after metabolism, a prediction of the metabolism first theory.

Unanswered questions include how life began to use DNA and RNA as a source of genetic information, a question that scientists are working hard to answer. It also remains unclear how life got its first membranes making it possible for them to leave their birthplace. Despite the difficulties of these questions, scientists will continue to address these questions. Admittedly, we may never find the exact answer of how life began on the Earth, but then again, we could have an answer that works well within a few years.

My favorite implication of the metabolic theory is that life originated as an extension of geological processes. Not only are we stardust, but our distant ancestors may have originated from a metabolically active rock in an ancient ocean!

All Modern Life Descended from the First Cells

Life started simple out of necessity. However, once life got started it has continually persevered for billions of years, evolving into all the plants and animals you see today. Today, there are two basic types of cells based on their structure, the simpler prokaryotic cells and the structurally more complex eukaryotic cells. The first cells on the planet were small prokaryotic cells mostly lacking internal structures. Today, cell theory predicts that the cell is the basic unit of life and since the origin of life, all cells have come from preexisting cells.

Today, all living organisms share many traits in common, evidence strongly supporting a single origin of life at least 3.8 billion years ago and potentially 100-200 million years earlier. By studying the similarities of living organisms today, we can predict the characteristics of our last universal common ancestor, affectionately known as LUCA. It works because, as life reproduces, genetic information is passed to the next generation. Although, life changes over time, some characteristics have remained little changed since the first living organisms.

Several features unite all life to a single origin, including ATP (adenosine triphosphate), the energy currency of life. By now, you may realize the importance of energy for cells. Ironically, cells can't directly use the potential energy in most organic molecules, including carbohydrates, for most of the work they need to do. Instead, cells must transfer energy to ATP, which they can use. Humans use about our body weight each day in ATP! In Chapter 3, I will discuss how our cells efficiently make ATP in cellular respiration. Importantly, it involves chemical reactions and differences in pH, much like the differences in pH between an alkaline vent and an acidic ocean.

Another compelling evidence for a single origin of life is that every living organism stores genetic information in DNA. Even more remarkable, the information to create bacteria, a tree, a fungus, an insect, or even a human is stored in a molecule made of just four different building blocks or the nucleotides, Adenine (A), Guanine (G) Thymine (T), and Cytosine (C). Imagine if the English language only used four letters to compose every song, story, poem, or text book. The story of life is written in the sequence of nucleotides forming DNA.

Since the 1960s, we have learned that the language of life, known as the genetic code, is basically the same in every living organism, inherited unchanged from a common ancestor that lived 3.8 billion years ago. Although we haven't shared an ancestor with bacteria for over 2 billion years, we still use the same genetic code. This means we can insert any human gene into bacteria and the bacteria would produce the same protein. In fact, that is how we make insulin to treat diabetes. Scientists isolated the human gene to make insulin, inserted it into a bacteria cell, grew it, and then harvested the insulin. In addition to having a universal genetic code, all living organisms share genes to make the same proteins to carry out similar functions. For example, every cell on the planet uses glycolysis, a series of 11 chemical reactions to break-down glucose into pyruvate.

In addition to DNA, cells use ribonucleic acid (RNA), a very close relative of DNA as its name would suggest. DNA has one job, it stores genetic information, but RNA carries the information from the DNA to ribosome where proteins are made from their amino acid building blocks. Not only does all life use ribosomes and RNA to make proteins, they also use the same 20 amino acids, building blocks of proteins. The next time you look out the window at a tree, just remember that you are made of the same building blocks, share the same genetic code, and actually share similar genes!

Lastly, cellular membranes made of similar materials surround all cells. They form on their own and create a barrier between the ordered, living component of the cells and the inanimate outside world. Life also uses their membranes to make ATP. Additional features shared by all life includes using metal catalysts to speed up chemical reactions. Today, many protein catalyst require a metal, like iron to make a reaction happen. Even hemoglobin uses atoms to transport oxygen. Based on the similarity of all cells today, we can predict that life had a single origin with certain characteristics still present in every living organism today.

The first cells likely emerged from chemical processes in a warm alkaline vent by using a hydrogen gradient to do work. They created ATP by using a natural gradient across a cellular membrane, used DNA to store genetic information, and used RNA to make proteins from the same twenty amino acids.

Chapter 3

Energy,
The Secret to Life

The "Secret of Life"
May well be chemiosmosis,
The ability to harness the potential energy
in proton gradients.

Life is an island of low entropy,
out of equilibrium with its environment

Introduction

In the Empire Strikes Back, during Luke's Jedi training, Yoda describes the Force as a mystical energy field that flows through life. Yoda was not far from reality, energy does flow through life. Cut life off from energy and it will die.

Energy is the secret to life, just like in the old sci-fi movies when the mad scientists proclaims, "I have discovered the secret to life" and they go on to a corpse hooked up to wires waiting for a lightening strike to supply the energy to reanimate life. This notion of energy bringing back life comes from observations in the 1800s that a small electrical current caused the muscles of dead animals to twitch.

All life uses energy to create order and without energy, life decays to equilibrium. To know why life needs energy, it's important to understand the nature of energy, the laws that govern it, and how life extracts energy from the environment, the major focus of this chapter.

Perhaps the best way to understand energy is to know what it does. Simply put, energy makes things happen. It is a property of matter governed by the laws of thermodynamics, which broadly state that energy cannot be created or destroyed. But, every time you use it, energy degrades, becoming less usable. We need energy constantly, it allows animals to be active and humans to maintain a very large brain. Therefore, it should be of no surprise that life has evolved efficient ways to extract energy from the environment, including chemiosmosis.

Life needs energy, but it can't use any type of energy. For example, simply adding heat won't work, if it did, you would only have to sit in the sun to power up. Instead life converts energy from the environment to a high-energy molecule called ATP, known as the energy currency of life. Cells use various processes to make ATP, but the most important is chemiosmosis. The reason for the universality of chemiosmosis across all life goes back to the first life on Earth harnessing natural sources of energy, perhaps from geological features like alkaline vents. Chemiosmosis makes the vast majority of ATP for every cell. It uses the flow of protons down their gradient to make ATP, like water flowing down hill and turning a water wheel to do work. Interestingly, the current theory of life's origins predict that life began in alkaline vents in an acidic ocean, creating a natural flow of protons that cells could use for energy.

Once life evolved the ability to harness a natural flow of protons to make ATP, the next step in evolution was to create their own proton gradients, enabling life to leave its birthplace. Once that happened, the structures were modified for photosynthesis and eventually aerobic respiration. Eventually, photosynthesis changed the planet paving the way for aerobic respiration making complex life possible.

The Nature of Energy

Energy is best defined by what it does; energy causes things to happen. It is a property of objects that can be transferred or transformed and is subject to the laws of thermodynamics. Energy allows you to effect some kind of change in the universe, or more commonly, it allows you to do work. To a physicist, work is the ability to move an object.

There are several ways we can measure energy; the most standard measurement is called the Joule, which is a measure of energy transferred to objects. However, in biological systems, the calorie is commonly used. One calorie is the energy required to raise one gram of water one degree Celsius. Whenever you read calories on food packaging, it is actually in kilocalories or Cal. One kcal of energy could raise a liter of water (about 1 quart) 1°C, or about 1.9°F.

Our understanding of energy was greatly expanded in 1905 when Einstein published his revolutionary Special Theory of Relativity producing perhaps the most famous equation in the world: $E=mc^2$. Einstein's insights regarding the nature energy and mass was revolutionary for the field of physics paving the way for the nuclear age of atomic weapons and nuclear reactors.

The relationship between mass and energy is explained by Einstein's equation $E=mc^2$ where E = energy, m = mass, and c = the speed of light, a staggering 186,282 miles per second (300,000 km/s). It means the mass of an object is also measure of its energy content. To determine the energy content of an object, you would measure its mass and multiply it by the speed of light squared. This means an enormous amount of energy is tied up in the mass of an object. When the U.S. exploded the first atom bomb in 1945 at the Trinity Site in New Mexico, approximately 1 gram of mass was converted to energy, mostly in the form of heat and light.

Energy released from a chemical reaction like a fire, converts a tiny amount of mass into energy. In our everyday world, whenever we use energy, the mass loss to energy is insignificant, measured in millionths to trillionths of a gram, so we won't worry about it here. A consequence of the relationship between mass and energy predicted by Special Theory of Relativity is that you cannot travel faster than the speed of light, it is a universal speed limit. The faster you go, the more energy you need. To go faster than the speed of light would require an infinite amount of energy to push an object past the speed of light.

To use an everyday example to understand why it takes more energy to keep going faster, imagine the difference in energy we spend walking versus running. Most of us can easily walk a mile, to do it in an hour would require little energy. To run a mile in 6 minutes requires you to not only be in good shape, but you also spend more energy to go the same distance. The faster you run, the more energy you use. It's also why your car gets worse gas mileage at higher speeds, it takes more energy to make your car move faster.

The energy and mass equivalence isn't just limited to physics or science fiction. In biology, animals require energy to move, the faster they run, swim, or fly, the more energy they use, placing physiological limits to how fast animals can move and the duration for how long they can maintain top speeds. It's easy to run a mile, but very few people can sprint a mile, and no one can sprint a marathon.

Recall that the laws of thermodynamics govern energy transformations. There are actually four laws of thermodynamics, but in biology, the first two laws matter the most for our purposes. Before diving into the laws of thermodynamics, keep in mind that systems are either open or closed. In a closed system, like the whole universe, nothing gets in or out. In contrast, the Earth, all its ecosystems, and living organisms are open systems where energy is constantly flowing through them. In almost every ecosystem, energy enters as sunlight and exits as heat.

The first law of thermodynamics tells us that in a closed system, energy is not created nor destroyed. For this reason, it is also called the conservation of energy because the total amount of energy is always conserved in a closed system. The first law also tells us that energy can be transferred from one object to another, and it can be transformed into different types of energy. For example, plants use photosynthesis to transform the kinetic energy in sunlight into the potential energy stored in carbohydrates.

Life being an open system, must maintain a constant flow of energy, or it will die. We can use ourselves as an example to understand energy transformations in living organisms. We obtain all our energy from the foods we eat, which contain potential energy. Our cellular metabolism transforms potential energy into kinetic energy when we go for a walk. The faster we walk the more energy we use. As we break down carbohydrates during exercise, the chemical reactions transform most of the potential energy in the foods we eat into heat, another form of kinetic energy. That's why we heat up and sweat when we exercise.

The second law of thermodynamics is nature's reality check for using energy. It states that every time energy is transferred or transformed, or anytime we use energy to do work, some of that energy becomes less usable as the total entropy increases in a closed system. There are a couple of ways to view entropy; the first is a measure of disorder or randomness. If you never clean your house, then entropy would continually increase as it becomes progressively messier. Second, as entropy increases, the amount of energy available to do work decreases.

Using energy releases heat to the environment increasing entropy. The more intensely you exercise, the more heat you generate and release. Unfortunately, heat is not a very usable form of energy; it is just the random motion of matter in a system. The more heat, the higher the entropy.

From a physics point of view, life is an island of low entropy out of equilibrium with its surroundings. Life uses a constant supply of energy to keep it more ordered than the chaos on the outside. If life is creating order and lowering entropy, then death is the exact opposite. Death is the decay to equilibrium, a state of high entropy.

There are Two Types of Energy

Kinetic energy is the energy of motion, the faster something is moving, the more energy it has. The air around you contains trillions upon trillions of tiny molecules of oxygen (O_2), nitrogen (N_2), carbon dioxide (CO_2) and other trace gases. These air molecules move fast, about 1000 miles per hour, which means they have kinetic energy. In fact, the air contains so much kinetic energy that it exerts about 15 pounds of pressure over every square inch of your body. All atoms and molecules have at least some kinetic energy even in the depths of space between galaxies. Known as thermal energy, we measure the average thermal energy of a system, such as your body, as temperature. The higher the temperature, the more kinetic energy.

In addition to thermal energy, there are other types of kinetic energy including electromagnetic radiation, which includes visible light, microwaves, radio waves, ultraviolet waves (UV), and X-rays. Electromagnetic radiation is made of photons, tiny particles that travel in waves at the speed of light. When you step into the sunlight, you can feel the energy as heat on your skin. Stay in the sun long enough and ultraviolet light will begin to damage your skin cells, accelerating the aging process. Other types of kinetic energy include sound waves and electricity. Sound waves form when something vibrates, but unlike light waves, sound waves cannot travel through a vacuum, including outer space (you would never hear a spaceship exploding in space!) When electrons move from one place to another, that makes electricity

Potential energy is any type of stored energy. As an animal, we extract energy from the foods we eat as part of our metabolism, which is the sum of all the chemical reactions in our body. Chemical reactions make new molecules by breaking and making bonds that hold elements together. It takes kinetic energy to break chemical bonds, which, in turn, releases energy when new chemical bonds are formed.

Ever watched a campfire's flames rising from the wood? To make a fire requires wood, oxygen, and a match. The wood supplies the fuel in the form of cellulose molecules made of glucose, the same sugar in our soft-drinks. Cellulose is made of three elements, carbon hydrogen, and oxygen held together by chemical bonds. The match supplies the energy to break the bonds holding the elements together, starting a chemical reaction. Once the bonds are broken, the elements quickly react with the oxygen in the atmosphere to make carbon dioxide and water. When these new bonds form, potential energy is transformed to kinetic energy releasing photons of energy that you see as light and feel as heat.

When molecule stores a lot of potential energy, the chemical bonds become less stable and are easily broken. For example, octane, a component of gasoline, contain lots of potential energy making it easy to break the bonds holding the molecule together. It only takes a small spark to provide enough energy to break the chemical bonds, igniting the gasoline. Cells use ATP, another high energy molecule, to temporarily store energy. In contrast, molecules such as water and carbon dioxide store very little energy because the chemical bonds holding them together are difficult to break, making these molecules stable. If you've ever thrown a match into water, it simply goes out. The flame lacks sufficient kinetic energy required to break apart the water molecule into its elements, oxygen (O) and hydrogen (H).

Similar to kinetic energy, potential energy comes in different forms. For example, gravitational energy works based on position, the higher you place an object, the more potential energy it has. If you were to drop a glass only a few inches from the floor, little would likely happen. Drop the same glass from a few a few feet and it will likely shatter. And of course, water stored in large lakes behind a dam is also a form of potential energy.

We learned from Einstein that an immense amount of energy is tied up in the mass of objects. Inside our sun, nuclear fusion slams hydrogen atoms together forming helium, and in the process, releases vast amounts of energy. However, in the last 4.6 billion years, the sun has only lost about 0.008% of its mass, or about the equivalent of Earth's mass. Nuclear energy is stored in the nucleus of atoms. Radioactive isotopes decay into smaller elements releasing kinetic energy. Nuclear reactors use the heat generated by radioactive decay to generate electricity.

Cells spend a lot of energy moving ions across their membranes. As they do so, one side becomes more positively charged than the other side forming a membrane potential, similar to how a battery stores energy. Because there is both an electrical gradient and a chemical gradient due to the different concentrations of ions across the cellular membrane, they are commonly called electrochemical gradients.

The energy stored as an electrochemical gradient across cellular membranes allows cells to regulate how much water is in the cell or the movement of other materials in and out of the cell. And, as I've stated earlier, cells can also use electrochemical gradients to make ATP. Based on the importance of membrane potentials to cells, something the first cells got their start by using the potential energy of natural electrochemical gradients found in thermal vents created by differences in pH.

Chemical Reactions
Transfer and Transform Energy

Life gets energy from its environment. Based on how they get energy, organisms can be broadly classified as autotrophs or heterotrophs. Although, all organisms break down organic molecules and transfer the energy to ATP, autotrophs fix carbon dioxide into the organic molecules they need by using energy from their environment such as light (photosynthesis), or potential energy from inorganic chemicals or minerals (chemosynthesis). Basically, autotrophs take energy from the environment and store it in organic molecules by adding hydrogen to carbon dioxide. In contrast, heterotrophs only use chemical energy stored in organic molecules made by autotrophs. As an animal, we are heterotrophs that obtain energy from the foods we eat.

Inside our cells, energy-rich organic molecules, including carbohydrates and fats, are broken down and their energy is used to do work within cells via ATP. We are most familiar with work as the ability to move an object. In addition to powering our muscles, life at the cellular level spends a lot energy moving objects across their membranes to maintain an internal environment that is different from outside the cell, known as homeostasis.

Every day, your body spends enormous amounts of energy pumping ions (mostly hydrogen, sodium, potassium, and calcium) across membranes creating electrochemical gradients to maintain homeostasis, regulate the amount of water inside cells, contract muscles for movement, or send nerve signals for rapid response. Cells also use energy to make larger molecules from smaller building blocks. All of these chemical reactions inside a cell, breaking molecules down and building larger ones, is our metabolism, which consists of both exergonic and endergonic reactions.

Exergonic Reactions: The energy cells use for movement, maintaining homeostasis, or building larger molecules, comes from exergonic reactions that release energy. Although the organic molecules in our food store energy, most are relatively stable at room temperature. They simply do not randomly burst into flames breaking down into smaller molecules. Don't worry, despite what you've seen on an Internet video, you cannot burst into flames by spontaneous combustion. To start any chemical reaction requires an input of energy to break the chemical bonds between the elements in a molecule. The more potential energy stored in a molecule, the less energy is required to start the reaction, it's the reason why you shouldn't smoke around gasoline. It won't spontaneously combust, but it doesn't take much energy to get it started.

Fire is a good example of an exergonic reaction. If you've ever started a fire with a match, you are supplying the energy required to break the chemical bonds between the elements in the cellulose. Once the original bonds are broken, the elements want to quickly form new bonds so they will be chemically stable. Once an exergonic reaction gets started, the energy released from making the products is enough to break the bonds of the reactants. This cycle continues until all the reactants are gone. That means, once you light a fire, it will continue to burn until it is out of fuel.

Below is the general chemical equation for cellular respiration, a series of chemical reaction in our cells that breaks down organic molecules and transfers the energy to ATP. The overall chemical reaction is also similar to the burning of wood or gasoline.

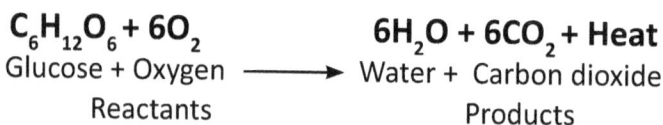

$$C_6H_{12}O_6 + 6O_2 \longrightarrow 6H_2O + 6CO_2 + Heat$$

Glucose + Oxygen $\longrightarrow$ Water + Carbon dioxide

Reactants   Products

In this reaction, there are different molecules on each side. The reactants on the left side include one molecule of glucose ($C_6H_{12}O_6$) and six molecules of oxygen (O_2). On the right side of the equation are the products, six molecules of carbon dioxide (CO_2) and six molecules of water (H_2O). It is also a balanced chemical equation because the same number of elements are on each side; for example, there are six carbon atoms in the reactants and in the products.

It's also important to know that chemical reactions make new molecules, but do not make new elements. If you have six molecules of oxygen in the reactants, then you will have six molecules of oxygen in the products, always. No matter how many chemical reactions take place, the elements will never be changed, like playing with Legos, no matter how many different space ships you make, you always keep building with the same bocks.

Recall that three subatomic particles form the elements, protons, neutrons and electrons and only the electrons form chemical bonds. Some elements attract electrons more than others and oxygen really likes electrons, more so than almost every other element. When organic molecules react with oxygen, they become oxidized, a type of chemical reaction that occurs when oxygen takes away electrons.

Oxidized molecules are often smaller and store less potential energy than reduced organic molecules. Looking back at our chemical reaction, glucose gets oxidized to carbon dioxide and water when it reacts with water. The oxidation of glucose and other organic molecules releases heat, and the products are simpler than the reactants. We breathe in oxygen to oxidize organic molecules, allowing our cells to extract energy from them. That's why we breathe in oxygen, breathe out carbon dioxide, and our bodies get warmer when we exercise. But, no matter how efficient a cell is at capturing energy, most is lost as heat to the environment, satisfying the second law of thermodynamics. If it weren't for that pesky law, our bodies could continuously recycle energy, similar to how the elements are recycled. However, it takes usable energy to keep those elements recycling.

Without an input of energy, or the presence of enzymes, exergonic reactions, while chemically favorable, may occur very slowly, or not at all. Because oxygen really likes electrons, every once in a while, it will react with sugar or other organic molecules to form water and carbon dioxide. But, the molecules are shy, they have to contact the glucose molecule in exactly the right way, or nothing happens. As a result, sugar sitting on your table lasts a long time. It would eventually break down, but it may take decades to do so. That's why cells use enzymes to speed up chemical reactions.

Endergonic Reactions: Endergonic require an input of energy and create order inside cells by building larger and more complex molecules. Cells use some of the energy from exergonic reactions to supply the energy necessary for building larger molecules. The main reason why an endergonic reaction requires an input of energy is because the simpler molecules forming the reactants have strong bonds that are hard to break. So, more energy is required to break the bonds of the reactants than energy is released when the products are formed. As a result, the products of an endergonic reaction contain more potential energy than the reactants.

Recall that exergonic reactions supply the energy required to build complex molecules. However, the energy released from the exergonic reaction must always be greater than the energy required for the endergonic reaction. Without a constant supply of energy coming into a cell, endergonic reactions would grind to a halt and the cell could no longer build complex molecules and eventually cell would decay as it reaches equilibrium with its environment. The most well-known endergonic reaction is photosynthesis, where plants use the energy in sunlight to fix carbon into sugars

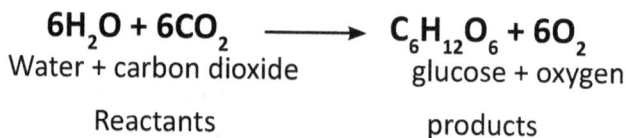

$$6H_2O + 6CO_2 \longrightarrow C_6H_{12}O_6 + 6O_2$$

Water + carbon dioxide	glucose + oxygen
Reactants	products

Photosynthesis requires lots of energy to break the strong chemical bonds in water and carbon dioxide. That's why plants rely on the energy in sunlight for this reaction. The energy in sunlight is used to rip the hydrogen atoms from the water and in the process release oxygen. Then, through a series of chemical reactions, hydrogen atoms are added to the carbon dioxide, making an organic molecule. Photosynthesis is one of the most important chemical reactions on the planet because it makes carbon available to animals, every carbon atom in your body was once in a molecule of carbon dioxide.

Making ATP: The Energy Currency of Life

All life uses the same energy currency, a high energy molecule called adenosine triphosphate, better known as ATP. The more ATP available inside a cell, the more work that can be done. Except for photosynthesis, energy entering a cell cannot be used directly by the cell, instead it is transferred to ATP. ATP is so important to cells, that some cellular biologist have called it the molecule of life. Every day, you use about your body's weight of ATP to carry out all of your body's functions from breathing, running, studying and storing memories, or making proteins in your cells. The universality of ATP among all life strongly supports a universal common ancestor.

A molecule of phosphate is comprised mostly of electron grabbing oxygen atoms attaining a slightly negative charge. With three phosphate groups attached to the adenosine molecule, the negative charges of the oxygens repel each other, storing lots of potential energy, and making the molecule somewhat unstable. When ATP is used, the last phosphate group comes off forming adenosine diphosphate or ADP.

The regeneration of ATP from the less energetic ADP is of paramount importance for cells and they have evolved two basic ways to accomplish this task, substrate level phosphorylation and chemiosmosis. Substrate level phosphorylation uses a protein enzyme to add a phosphate group to ADP, forming ATP. Because the process requires energy, it is directly connected to the energy released from the break-down of organic molecules. It is an example of energy coupling where cells transfer the energy released from one reaction to another.

Perhaps the most commonly used source of energy for substrate level phosphorylation comes from a 10-step metabolic process called glycolysis, which literally means 'splitting sugar'. It is the first stage of a much larger series of chemical reactions known as cellular respiration, which I will discuss in more detail later in this chapter. Because nearly every cell produces some ATP through glycolysis, it is thought to have evolved early in the history of life, further supporting of a single-common ancestor for all life.

Despite the fact that most organisms make ATP by glycolysis, it is not the most efficient method. For example, it takes an exact amount of energy to make ATP, about 7.3kcal/mol. If an exergonic reaction releases 12.3 kcal/mol, then the extra energy is lost to the environment as heat. Or, if you don't have enough energy, you don't make any ATP, and the energy simply lost to the environment. A 'mol' is an abbreviation for a mole where one mole of an element would weigh its atomic weight in grams. Because atoms are tiny, it takes a lot of them to equal their weight in grams, in fact about 6.023×10^{23}, or a little over 600 billion trillion atoms of carbon-12 to weigh 12 grams.

Another way to better understand the reason why combining two reactions together to make ATP is not very efficient is to imagine purchasing a soft-drink that cost $1.50/20 ounces. If you put in $5 a vending machine, you get one soda, but you lost your $3.50 in change. On the other hand, if you put in $1.49 in change, no soda for you, and you lose the money you put into the vending machine! The same principle applies for making ATP. If a reaction releases 7.2kcal/mole of energy, you do not get any ATP, that energy is simply lost to the environment.

Chemiosmosis: The Secret of Life

If you've ever wondered if there is a "secret to life", it could be chemiosmosis; the ability to use proton gradients to efficiently make ATP. Chemiosmosis literally means "to push". It is nature at its most efficient, harnessing the potential energy stored in electrochemical gradients across a cell's membrane.

One way to understand gradients is that they are a change in some value across a distance. If you've ever walked up a hill, then you have gone up an elevational gradient, where the elevation changes over the distance you walked. An electrochemical gradient has two parts.

Chemisosmosis

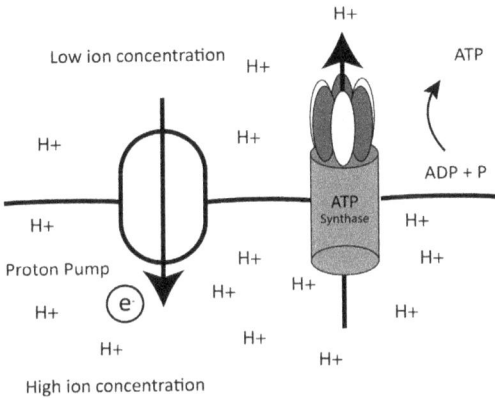

Chemiosmosis is an efficient way of making ATP by using the potential energy found in proton gradients.

Different charges across the membrane forms the electrical gradient, like a battery stores potential energy as a voltage. Whereas the difference in ion concentrations across a membrane forms the chemical gradient.

More protons on one side of a membrane forms a chemical gradient. Similarly, more positively charged protons on one side of a membrane forms an electrical gradient because one side becomes more positively charged. Combining the electrical and chemical gradients makes an electrochemical gradient, which stores potential energy that cells can use.

The universe does not like to be out of equilibrium. Because more protons are on one side of the membrane, the cell is out of equilibrium. However, the protons want to be in equilibrium with equal concentrations on each side of the membrane. But, the ions can't easily cross a cellular membrane, which restricts their movements. It is similar to water behind a dam, the water can't move through the concrete, so it builds up to higher levels behind the dam, storing energy.

To use potential energy stored in the electrochemical gradient, the ancestors to all cells evolved a large protein complex called ATP synthase. The 'ase' ending means this is an enzyme that helps a chemical reaction take place. When lots of protons build up on one side of a membrane, they "push" their way through the ATP synthase to the other side of the membrane in an effort to reach equilibrium. As the protons push their way down their electrochemical gradient, the ATP synthase uses the kinetic energy of the moving protons to make ATP.

Chemiosmosis efficiently makes ATP because energy can be stored in a gradient across a membrane until there's enough to make a molecule of ATP. It's like putting your change in a piggy bank, once you have $1.50, you can buy your soda from the vending machine and you don't waste your money.

Based on our current theories of life's origins and early evolution, the next major evolutionary leap for life occurred when cells began to generate their own proton gradients, not relying on the natural gradients in the alkaline vents. This crucial step allowed cells to live independently of their place of origins and spread to new areas and exploit new resources, such as the energy in sunlight.

The answer for these cells was the evolution of large protein pumps powered by electrons that actively moved protons across cellular membranes, generating their proton (electrochemical) gradients. The first organisms that generated their own proton gradients likely obtained their energy and electrons from various sources in their environment. In the earliest autotrophs, or self-feeding organisms, the source of electrons likely came from minerals or rocks. Whereas, the first heterotrophs obtained their electrons from organic molecules. While there are many sources of electrons and energy to power the proton pumps, there must also have been electron acceptors that removed electrons, or the whole process would have stopped as it reached equilibrium.

Today the series of proton pumps and electron carriers is collectively called the electron transport chain. Together with ATP synthase the entire process efficiently makes ATP through chemiosmosis, a process that is ubiquitous among all living cells. Overtime, the electron transport chain has evolved to use different sources of electrons, energy, and electron acceptors.

If ATP is the molecule of life, then chemiosmosis is one of the most crucial metabolic processes of life. Based on the ubiquitous nature of chemiosmosis, it's generally accepted that for more than 3.8 billion years, some form of chemiosmosis has been preserved in every cell, inherited from the first living organisms on the planet. In the next two sections, I explain how cells have re-purposed the molecular machinery of chemiosmosis to use new sources of energy and electrons in photosynthesis and a new electron acceptor in aerobic respiration.

Photosynthesis
Using an Abundant Energy Source

The evolution of photosynthesis was one of the most important innovations in the history of life. Photosynthesis means 'putting together' (synthesis) with 'light' (photo). It takes the energy in sunlight and stores it in organic molecules, mainly carbohydrates. Small prokaryotes called cyanobacteria were most likely the first group of organisms to evolve photosynthesis perhaps some 3.5 billion years ago. Most autotrophs use photosynthesis to store energy in organic molecules by using energy from the sun to form organic molecules from carbon dioxide.

Photosynthesis is a remarkable evolutionary innovation for two reasons. First, it gave life access to a practically unlimited source of energy from the sun, no longer relying on energy from natural proton gradients or from acquiring organic molecules in their environment. Second, these ancient cyanobacteria used the energy from the sun to create all the organic molecules they would ever need. That's why photosynthetic organisms are at the base of almost every ecosystem in the world; they make carbon and energy readily available to the rest of us.

I can't understate the importance of photosynthesis; just think, every carbon atom in your body was once part of a carbon dioxide molecule in the atmosphere fixed into an organic molecule through photosynthesis. Therefore, photosynthetic organisms form the foundation of almost every ecosystem.

The chemical equation for photosynthesis may seem simple, but in reality, it is a multi-step process requiring sunlight, water, carbon dioxide, many proteins, and cellular structures dedicated to the process. There are two main parts of photosynthesis, the light reaction and the synthesis reaction, also known as the Calvin cycle. Typically, we learn photosynthesis as it takes place in the chloroplasts inside plant cells. Chloroplast are tiny organelles, or tiny cellular organs that were once free-living bacteria. Plants never evolved the ability to perform photosynthesis, instead, the ancestor of all plants actually acquired the ability for photosynthesis by engulfing cyanobacteria. Over time, they evolved into the chloroplasts that are found in plant cells.

What is Visible Light?

Before we dive further into photosynthesis, it's good to know a little about the properties of light. Visible light is a small part of the electromagnetic spectrum formed by tiny packets of energy called photons that travel in waves. The shorter the wavelength, the more energy in the photons. We see a tiny fraction of the electromagnetic spectrum as visible light. For example, the wavelength of red light is about 700 nanometers and blue light is about 380 nanometers. Ultraviolet light has a wavelength ranging from 100-380 nanometers and carries enough energy to damage your skin. Infrared light that we can feel as heat has a wavelength from about 800 nanometers to about 1 millimeter. A nanometer is one billionth of a meter. To put that tiny size into perspective; if you were to enlarge an object that was one nanometer to a millimeter in length, (about the width of the letters you are reading) it would be the same as making a meter stick one thousand kilometers (about 600 miles) long!

The Light Reaction

How do plants transform the energy in sunlight in **the Light Reaction?** Whether or not photosynthesis takes place in a cyanobacteria or in the chloroplasts of a plant cell, the basics of photosynthesis are similar. It begins with the light reaction, which accomplishes several tasks. Importantly, it transforms the energy in sunlight into a usable form of chemical energy in ATP through chemiosmosis, much like I just described.

How does the energy in light waves get temporarily stored in ATP during photosynthesis? In the light reaction, the proton gradient required for chemiosmosis is created by using electrons to power proton pumps as they travel down an electron transport chain. Does this sound familiar? The transformation of light energy to ATP in the light reactions is very similar to how the first cells made ATP by using proton gradients and chemiosmosis. Remarkably, it was probably not a giant evolutionary leap for life to evolve the light reaction of photosynthesis. A recurrent theme in evolution is that structures are continually re-purposed for new tasks. In the case of the light reaction, electron transport chains were re-purposed to help transform light energy to ATP. Although the source of electrons and energy is different for the light reaction.

Inside photosynthetic cells, the light reaction takes place in large molecular complexes called photosystems embedded in cellular membranes dedicated to the light reaction (you can think of a photosystem as a large satellite dish to collect light). Its purpose is two-fold; first to use the energy in sunlight to split water, ripping the electrons away from oxygen. The second purpose is to elevate the electrons in energy to power an electron transport chain that generates a proton gradient. A second photosystem re-energizes the electrons after they have lost their energy traveling through the electron transport chain and then get used in the Calvin cycle. Let's take a closer look.

The light reaction begins when chlorophyll, the pigment that makes plants green, absorbs and concentrates light energy. Once the light energy has been "captured" in the chlorophyll, some of that energy is used to split water into hydrogen and oxygen. The oxygen exits the cell as free oxygen (O_2), an important consequence of photosynthesis that I will return to later in this chapter. Water is tough to split apart because oxygen likes to hold onto its electrons. Therefore, plants need the energy in sunlight to split water by pulling electrons away from oxygen. Also, the electrons that came from the splitting of water lack energy and at this point, can't be used to power the proton pumps.

The second purpose of the photosystem is to use light energy to excite the electrons in energy. Once the electrons have been elevated in energy, they go through a series of electron carriers and proton pumps collectively called the electron transport chain that creates an electrochemical gradient by pumping protons across a membrane, ultimately storing energy.

Embedded in the same membrane that houses the photosystems and the electron transport chains are ATP synthases. To reach equilibrium, the protons "push" their way through the ATP synthase and in the process, ATP is generated through chemiosmosis. Recall that this is the same process of chemiosmosis used to generate ATP in practically every cell on the planet. The energy in sunlight has now been transformed to potential energy stored in ATP, which will be used as an energy source for the Calvin cycle. After the electrons have traveled down the electron transport chain, they have once again lost their energy. At this point, a second photosystem energizes the electrons so they can be used to fix carbon dioxide into organic molecules in the Calvin cycle.

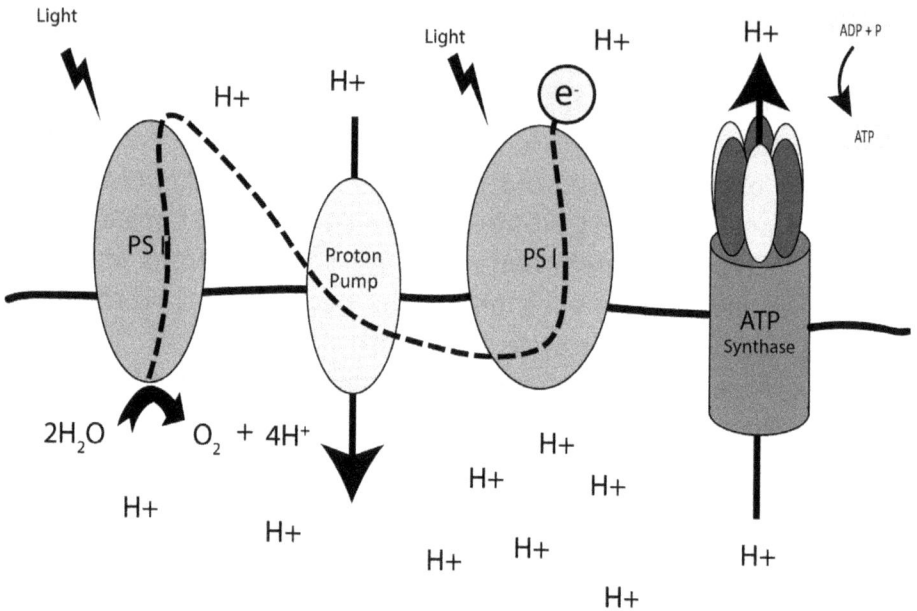

Photosynthesis uses the energy in sunlight to remove electrons from water and elevate them in energy so they can power proton pumps used to create a proton gradient. The potential energy is used by ATP synthase to make ATP through chemiosmosis.

To recap the light reaction, it begins when sunlight strikes the chlorophyll pigments in a photosystem, the energy is concentrated as water is split and electrons are energized. These high energy electrons are used to power proton pumps, generating the proton gradient used to make ATP through chemiosmosis. A second photosystem, re-energizes the electrons so they can be used to make organic molecules in the Calvin cycle.

The Calvin Cycle

The Calvin cycle is a series of chemical reactions that takes carbon dioxide (CO_2) and fixes it to organic molecules by adding hydrogen atoms to it. Carbon dioxide is very stable, or inert, it doesn't readily react with other molecules. It takes a lot of energy to make an organic molecule out of carbon dioxide because it's difficult to break the chemical bonds between carbon and oxygen. The energy for the Calvin cycle comes directly from energized electrons from the light reaction and the ATP produced by chemiosmosis. The source of hydrogens and electrons comes from the water that was split in the light reaction. Once carbon has been fixed, it becomes available to life. In fact, every carbon atom in your body was fixed in the Calvin cycle.

Aerobic Respiration
An Evolutionary Response to Free-Oxygen

The free-oxygen released from photosynthesis energized life on the planet. Sometime between 1.5 and 2 billion years ago, atmospheric oxygen levels reached a critical point, it became toxic to the life forms that had existed in a relatively oxygen-free environment for almost two billion years. A few researchers have gone so far as to call the rise in oxygen at this time the Great Oxygen Catastrophe because they speculate that the rising oxygen concentrations may have caused a mass extinction of microbial life not adapted to oxygen. Even today, bacteria known as anaerobes will die in the presence of oxygen as they have no defenses against its oxidizing tendencies. In reality, it probably was not so much of a catastrophe for life as it was an evolutionary opportunity.

Life has an uncanny ability to evolve and adapt to its surroundings, allowing it to exploit new resources, including oxygen. To use the energy locked up in organic molecules, heterotrophs must transfer it to ATP. There are several ways cells can do this, and one that they all use is chemiosmosis, similar to what takes place in photosynthesis and in the first cells. In a remarkable twist of fate, the same electron grabbing property that makes oxygen damaging to cells can be used to greatly improve the efficiency of chemiosmosis, making it possible to make much more ATP from the same amount of organic molecules.

When oxygen became abundant, some prokaryotic cells evolved a small modification to their electron transport chains allowing them to use oxygen as the final electron acceptor. This one small evolutionary step was a major leap forward for life. First, it added the protons that were used in chemiosmosis to the oxygen molecules and formed water; in one simple step the toxic effects of oxygen were neutralized. But, the real leap comes from oxygen's higher affinity for electrons than all the other electron acceptors previously used by life. This meant that the electrons could be used to do more work because of the bigger energy drop they could undergo. This larger energy drop generated bigger proton gradients allowing cells to extract much more energy from organic molecules, thus producing much more ATP from every organic molecule they consumed. Cells now had more energy to grow larger and become more active.

Based on molecular evidence, the first aerobically respiring bacteria were proteobacteria, a large group of bacteria including pathogenic species such *Salmonella* and *Helicobacter*. They also include the familiar *E. coli* found in our guts and some bacteria that fix nitrogen. Eventually, an aerobically respiring bacteria merged with another prokaryote and together they would go on to evolve into eukaryotes, a totally new type of cell. Inside almost all eukaryotes are the descendants of the proteobacteria, the mitochondria. They evolved into energy-converting organelle responsible for aerobic respiration and the vast majority of ATP production in all plants and animals. The evolution of eukaryotes is known as endosymbiosis and I will discuss in Chapter 7.

In our cells and all other eukaryotes, aerobic respiration uses three steps; glycolysis, Krebs cycle, and oxidative phosphorylation, which includes the electron transport chain and chemiosmosis. The first step, glycolysis does not take place inside the mitochondria. However, it's important to note that glycolysis and the Krebs cycle are at the center of cellular metabolism.

Glycolysis means splitting (lysis) sugar (glyco) in half. It is a catabolic pathway requiring 10 steps, with each step using a different protein enzyme. Glycolysis occurs in almost every cell, both prokaryotes and eukaryotes, on the planet and may be one of the oldest metabolic pathways because it does not require oxygen or specialized structures inside the cell. It only produces a net of 2 ATPs for every molecule of glucose, not very much ATP considering the amount of potential energy in sugars. Another fascinating aspect of glycolysis is that its end product, called pyruvate, can serve as a precursor for many other molecules including amino acids, lipids, and nucleotides; all building blocks of life.

The Krebs cycle, a series of 8 steps forced to be blindly memorized in introductory biology courses without any explanation of its importance, is at the heart of cellular respiration. It was named after Hans Krebs who spent ten years working out the steps of this cycle in the 1930s. Like glycolysis, it occurs in most cells, but we usually learn about it inside of mitochondria, which are only found in eukaryotic cells. In cellular respiration, the main purpose of the Krebs cycle is to break down organic molecules by stripping them of their hydrogens and electrons. Although this cycle only produces 2 ATP, similar to glycolysis, the electrons carry lots of energy needed for the final step of aerobic respiration. By the end of the Krebs cycle, the carbon atoms that entered are released as carbon dioxide that simply diffuse out of the cell and you eventually exhale them as you breathe.

Oxidative phosphorylation is the final step of aerobic respiration and it occurs on a membrane similar to photosynthesis. The high energy electrons stripped form organic molecules get transferred to an electron transport chain used to power large proton pumps to generate an electrochemical gradient. Once the gradient stores sufficient energy, the protons flow through ATP synthase to produce ATP by chemiosmosis. Sound familiar, this is essentially the same process we encountered in the light reaction of photosynthesis.

Inside the electron transport chain, the electrons flow from one protein complex to the next, doing work at each step. By the end of the electron transport chain, the electrons have lost their energy making it hard to remove them. Because oxygen really likes electrons, it pulls them off the electron transport chain, ensuring a continuous flow of electrons, ultimately, maintaining a proton gradient for chemiosmosis. Because the electrons all end up on oxygen at the end of aerobic respiration, we say that oxygen is the final electron acceptor. Not only does oxygen pick up the electrons, it goes on to react with the protons coming through the ATP synthase forming water.

Does the process of oxidative phosphorylation sound familiar? Here is the third example of ATP being produced by chemiosmosis, a process utilizing a proton gradient generated by an electron transport chain and protons flowing through ATP synthase. In fact, the ATP synthase found in cyanobacteria is similar to the ATP synthase found in the mitochondria of all eukaryotes. As you can see, chemiosmosis has been re-purposed several times over the course of evolutionary history. In photosynthesis it was re-purposed to transform energy from sunlight in ATP. In aerobic respiration, it was re-purposed to protect cells from the oxidizing effects of oxygen with the added benefit of producing more ATP.

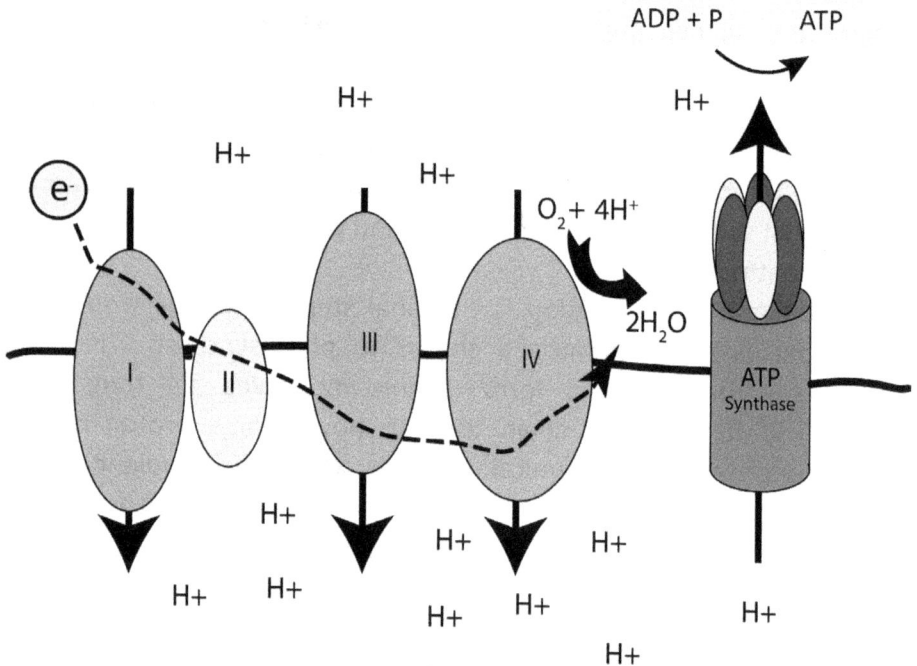

Oxidative phosphorylation is similar to photosynthesis, they both use electron transport chains to generate a proton gradient that powers the production of ATP as protons flow through an ATP synthase. Organic molecules serve as both a source of energy and electrons for the process.

Organisms that use anaerobic respiration don't use oxygen and produce far less ATP per organic molecule. This inefficiency ultimately places limits on the size and complexity of organisms. Without oxygen, larger more complex eukaryotic cells would never evolve. Life would have remained as simple unicellular organisms.

Life Changed the Earth Forever

Photosynthesis changed the world in many ways. To begin, every carbon atom in your body was once in a molecule of CO_2 in the atmosphere and was fixed into an organic molecule by the Calvin cycle. In fact, adding hydrogen atoms to carbon atoms to form organic molecules is fundamental to the existence of life; every organic molecule contains carbon and some hydrogen. To date, there are only about 4 metabolic pathways, including photosynthesis, that perform such an important chemical feat. Three of those are limited mostly to unique geological features or are not capable of growing very fast. If photosynthesis had never evolved, there probably wouldn't be the abundance of life on the Earth we see today.

Photosynthesis had another major effect on life. The byproduct of photosynthesis is free oxygen (O_2), a very reactive molecule. Oxygen as an element was abundant on the early Earth, but it was bound up in other molecules such as carbon dioxide. There was no free-oxygen in the atmosphere because it would quickly react with other molecules whether they were in the water, rocks, or newly formed minerals.

Over geological time, photosynthesis continually pumped out free-oxygen, eventually oxidizing almost all the iron in the oceans forming banded iron formations. Once the iron was removed from the oceans, they became the blue color we know today. It's hard to imagine, but the world looks much the way it does today due to oxygen in the atmosphere, a byproduct of photosynthesis! Additionally, we mine those ancient banded iron formations to make steel.

Oxygen had a huge impact on the atmosphere, the oceans, and every rock that has formed on the surface in the last 2 billion years. In fact, if scientists were to find an exoplanet with more than 1% oxygen in its atmosphere, it would be a smoking gun for photosynthesis, which is a biological process and clear indication of extraterrestrial life.

Another major consequence of atmospheric oxygen is that it reacts with ultraviolet light to form ozone (O_3). Ozone, even more reactive than oxygen, is a pollutant at the surface. But, high in the atmosphere, it blocks incoming ultraviolet light, protecting the surface from these damaging high-energy light waves. Without the thin layer of ozone, life could not exist on the surface, it would be killed by the ultraviolet light. Without ozone, life would be relegated to watery environments where it would be protected from harsh ultraviolet light.

Sometime, about 2 to 1.5 billion years ago, the first eukaryotic cells evolved in response to the higher levels of oxygen. The ensuing evolution of eukaryotic cells was the largest restructuring of cells in the history of the planet. This new type of cell was larger, had many mitochondria efficiently producing ATP, a cell nucleus that housed the DNA, and the insides became additionally compartmentalized into organelles. In fact, the name eukaryote means "new kernel" because these cells contained a nucleus unlike the smaller prokaryotes.

Eukaryotic cells evolved through a process called endosymbiosis. We don't know exactly how it happened, but about 2 billion years two prokaryotes merged to form one cell. The smaller prokaryote was an aerobically respiring bacteria using oxygen to efficiently make more ATP. The larger prokaryote may have been a methane producing Archaea. Eventually the bacteria evolved into mitochondria and the overall size of the cell grew as various organelles evolved compartmentalize the functions of eukaryotic cells.

About a billion years ago another round of endosymbiosis occurred when a eukaryote engulfed a cyanobacteria, which evolved into a chloroplast. This new cell had both mitochondria and chloroplasts, meaning they were autotrophs using photosynthesis and cellular respiration.

Eventually, some eukaryotes began forming colonies eventually evolving into the first multicellular organisms about 600 million years ago. One group went on to form plants, another into fungus, and of course, another into animals. Just think, life had been around for 3 billion years before the arrival of animal life! It would take another 150 million years for animals to begin colonizing the land. It appears that plants, fungus, and animals all began colonizing the land as early as 500-450 million years ago. By this time the Earth was already 4 billion years old.

Today, all multicellular organisms including plants, fungus, and animals are comprised of eukaryotes descended from those two prokaryotic cells that came together evolving into one cell. Prokaryotic cells never evolved into multicellular organisms and perhaps never will. But, they did evolve photosynthesis and aerobic respiration, neither plants nor animals evolved these abilities.

By now, I hope that you see the connections between energy and life. Life uses energy to create order, a tiny island of low entropy out of equilibrium with the universe. Understanding how life gets energy from the environment is important for understanding how cells work, how

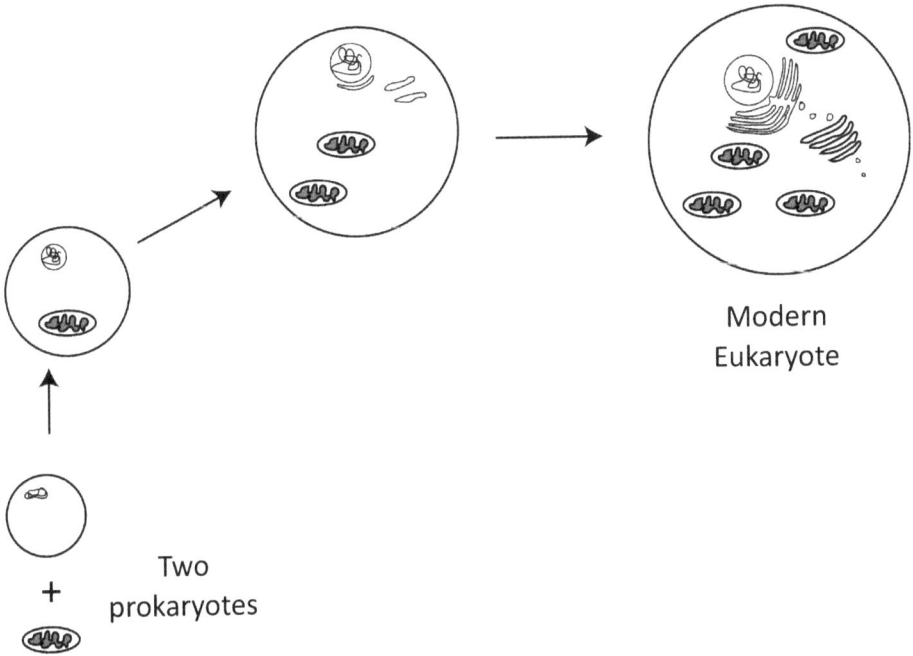

Modern
Eukaryote

Two
prokaryotes

+

Eukaryotic cells evolved through endosymbiosis when two prokaryotic cells merged and began living together as a single cell. Contrary to most popular illustrations, endosymbiosis did not occur when a "proto-eukaryote" engulfed a smaller bacteria.

life evolved, or how we might recognize alien life.

Endosymbiosis jump-started the evolution of eukaryotic cells. It occurred when two prokaryotes merged to form a single cell. Contrary to most illustrations in text books and on the web, almost certainly did not involve one cell engulfing or eating another cell.

Chapter 4

Scientific Revolutions: Evolution by Natural Selection

To raise new questions, new possibilities, to regard old problems from a new angle, requires a creative imagination and marks real advance in science.

Albert Einstein

To deny scientific discoveries is to deny reality. Nothing in biology makes sense except in the light of evolution

Theodosius Dobzhansky

Introduction

Scientific discoveries of the past few hundred years have revolutionized our understanding of how the natural world works. There have been many debates over which discoveries were the most important, were the most ground-breaking, or made the biggest contribution to the advancement of human understanding. I'm not going to weigh in on that debate because major advancements were all made by incredibly knowledgeable scientists who dedicated their lives to the advancement of knowledge. They were all intellectually gifted with imagination and creativity allowing them to elucidate the workings of the natural world.

At its heart, this chapter is about evolution by natural selection, a scientific advancement that like Newton's *Principia* or Einstein's theories of relativity, revolutionized the way we understand the natural world and conduct science. The impact of Darwin's theory reaches well beyond biology to include all fields of science. Upon its publication in 1859, it fundamentally changed the way we do science, forever removing any need to invoke supernatural explanations to understand the natural world. By the 1940s, evolution was firmly established as the central paradigm to biology.

Surprisingly, biology is the only major field of science that a large portion of the general public does not believe in its central paradigm. In fact, many revolutionary theories are met with resistance from other scientists, philosophers, and the general public because they challenge our deeply held beliefs, anecdotal knowledge, and often common-sense knowledge. Always keep in mind that science is a process about learning how the natural world works, it's based on evidence from experiments and observations. Science moves forward as we gain knowledge; hypotheses are proposed, tested, discarded, modified, or accepted. Over time, we build knowledge based on the works of many people, and our understanding of the natural world continues to grow. Sometimes we are wrong, but the repetitive nature of science makes it self-correcting as we home in on the right answers.

Scientific Revolutions

Revolutionary theories can take decades to be accepted by other scientists and even longer for society. That's because some theories are just too revolutionary based on the current knowledge and the *zeitgeist* of the time. *Zeitgeist* is a German word that means the 'spirit of the times', or the intellectual fashion of the time. Beginning with the Renaissance, philosophers and gentlemen naturalist began relying more on observations and experiments to advance science, rather than just accepting the conventional wisdom of their times. However, their early scientific discoveries often contradicted common sense knowledge and other beliefs held by society, making it difficult for people to accept their new findings because it directly challenged the accepted world view.

One example of a scientific discovery that defied common sense knowledge of the time was that the Earth is round. For most of humanity's existence, it was thought that the Earth was flat and the universe revolves around the Earth, giving credence to an inflated importance of man's place in the universe. This world view was based on observations that the sun moves across the sky and the Earth doesn't seem to move. Additionally, the surface of the Earth appears flat from where you are standing.

The first serious challenge to the flat Earth hypothesis dates back over 2250 years ago to the Greek Philosopher Eratosthenes who predicted that the Earth was round. He used an observation that two poles in different locations, one in Alexandria and the other in Syene, cast shadows of different lengths at the same time of day. Using basic geometry, Eratosthenes correctly calculated the size of the Earth to within a few hundred miles, which was a very accurate measurement.

Eratosthenes work was largely ignored for 17 centuries because it contradicted common sense knowledge when everyone thought the world was flat. He was finally proven correct in 1521 when the crew of Ferdinand Magellan successfully sailed around the Earth, proving once and for all it was round. But, people still thought the Earth was the center of the Universe. However, in 1543 a major challenge to this view was put forth when Copernicus published his theory of heliocentrism, predicting that the Earth and planets revolve around the sun. Once again, few people accepted his work because it was such a radical departure from the prevailing view of society placing the Earth at the center of the universe. Copernican heliocentrism eventually became widely accepted with additional discoveries a century later from Galileo.

Using the first telescope, Galileo observed moons orbiting Jupiter, lending more support to the notion that everything in the universe does not revolve around the Earth. The importance of Copernicus's and Galileo's heliocentric theory was that they used observations to develop testable models of the solar system, ultimately advancing our world view. Their heliocentric theory was extremely controversial because it removed the Earth as the center of the universe, reducing the view of man's place and importance in the universe. Today, we have satellites that routinely take images of the Earth against the vastness of the universe reminding us that the Earth is just a speck of dust in the vastness of the universe. The long-held belief of society based on "common sense knowledge" that the world is flat, or the Earth is the center of the universe has been totally disproved by the relentless progress of science.

Another scientific revolution occurred in the 1680s when Newton published his laws of motion and gravitation in one of the most important scientific books in European history, *Principia*. Newton's laws of gravity and motion were the first forces of nature to be clearly articulated through a mathematical treatise. In these laws, he described the effect of gravity on an object by stating that any object dropped on the Earth would fall at the same rate of 9.8 m/s^2. The same laws of gravity and motion were also used to describe the orbits of the planets beyond the Earth. By doing so, Newton clearly demonstrated that the fundamental forces of nature are the same everywhere. This thinking was also in stark contrast to the prevailing thoughts of the day in Europe that stemmed from Aristotle believing that the laws of nature were different beyond the moon, or the celestial realm.

Newton ushered in a new era of physics by making the characterizing of fundamental forces a priority of physics. Newtonian physics of gravitation and motion became the dominant paradigm of physics for the next 200 years. Although, Newton didn't explain gravity itself; that would take another two centuries and a profound intellectual leap in imagination and creativity by another great mind. Newton's laws were so successful that by the 1890s, some physicists were starting to believe that most of the laws of physics were about to be discovered. That naive world view was shattered when a relatively unknown physicist named Albert Einstein developed his revolutionary theories of relativity.

In 1905 Einstein published his Special Theory of Relativity describing the mass-energy equivalence in his famous equation $E=mc^2$. This theory began the atomic age and forever changed our perception of space and time. Perhaps, the most difficult aspect of special relativity is that it challenges our "common-sense" view of the universe by stating that there is no standard reference point to the universe because the speed of light is "fixed".

Here's another way to explain this, imagine you are standing on the road and a car passes by at 50 mph, if someone from the car throws a ball at 20 miles an hour, then the ball would be traveling at 70 mph relative to the person standing on the roadside. Einstein's remarkable insight was that this common-sense approach does not work for light; rather, he showed that light will always travel at the same speed, 186,282 miles per second (mps), relative to the observer. Meaning, if I turn on a flashlight, the speed of light will be 186,282 mps. If a car passes me at 100 mph and turns on the headlights, the light will be traveling at 186,282 mps for the car and for someone standing on the roadside! The speed of light is constant regardless of the relative motion of the source.

In 1915, Einstein once again devised another revolutionary theory, the General Theory of relativity, which explains gravity. The origins of General Relativity date back to Newton's Law of Gravity, which uses math to characterize the effects of gravity on objects including the planets, but did not explain how gravity worked. Cracks in Newton's laws of gravity were present in the early 1900s. Using Newton's laws of gravity, the orbits of the planets were pretty well worked out; except for Mercury, the closest planet to the sun. It turns out that its orbit was not exactly as it should be based solely on Newtonian physics. This was an indication that Newton's Law of Gravity may not be the whole story to gravity.

Einstein's General Theory of Relativity predicted that massive objects, like the sun or the Earth, curve space-time, creating a "valley" in the fabric of the universe that objects would fall into. Taking into account the curvature of space-time by the sun, Einstein's General Theory of relativity correctly predicted Mercury's orbit. Another verification of his theory came in 1919 when astronomers were able to verify that light from a distant star was bent as it passed by the sun. This independent verification of Einstein's theory made him an instant household name. However, it would take several more decades for the full impact of General Relativity to be appreciated because it also predicted black holes, the Big Bang, and an expanding universe before there was a single observation supporting any of these predictions.

Over the past 100 years, Generally Relativity has been repeatedly verified by numerous observations and experiments. Some physicists believe that Einstein's General Theory of Relativity may have been the biggest leap of scientific imagination in history. Surprisingly, Einstein never received a Nobel Prize for either of his theories of relativity.

Other discoveries have had a profound influence on our understanding of the world. In the 1800s, the laws of thermodynamics were developed to build steam engines, improving our understanding of energy. Maxwell's equations describe electric fields showing that light is not formed from some ethereal force, but is made by manipulating electric charges. In the 1920s, quantum mechanics explained the bizarre universe at the tiny scale of atoms where an electron can appear to be in two places at once, or where subatomic particles can instantaneously influence each other over great distances, a process known as quantum entanglement. Although this may sound of little practical use, our knowledge of quantum mechanics makes transistors used for modern computers possible. The Big Bang theory that grew out of Einstein's equations and Hubble's observations was thought to be so absurd by one famous astronomer that he called it the "Big Bang" to ridicule it.

Each one of these discoveries was born out of observations, experiments, imagination, and creativity. They have been repeatedly confirmed through extensive experiments and observations. It takes a lot of evidence and support to create a good scientific theory. In fact, the repetitive self-corrective nature of science either modifies or even discards theories if they do not hold up to the evidence. And lastly, the imagination and creativity of these scientists were vital in providing new insights into the natural world or creating new methods to test theories. Einstein, perhaps one of the greatest minds to have ever existed in the world, once proclaimed that imagination is more important than knowledge. Indeed, Einstein was special case considering he devised his revolutionary theories as thought experiments while working as a patent clerk!

A lesson from Einstein's Theories of Relativity, Copernicus's heliocentrism, and especially the strange behavior of tiny subatomic particles explained by quantum mechanics is that the natural world often defies our "common sense" views. How we commonly perceive the day-to-day world may not be how it actually works. These discoveries should be a reminder that we must use the results of our observations and experiments to understand the natural world and not rely solely on what we think is true.

Darwin's theory of evolution by natural selection was a revolutionary theory met with skepticism or outright denial. For many people the fact that species change over time defies common sense knowledge based on a very limited time frame. The problem stems from the fact that our lives are much too short to witness the large evolutionary changes in species taking place over millions of years. The events of our lives unfold in days, years, and even decades, and the entirety of human civilization is only a few thousands of years old. So short are our lives that we are ill-equipped to fully understand geological timescales of millions of years, let alone a 100 million years, and a billion years is basically unfathomable. In that amount of time, a lot of evolutionary changes can take place.

To put into perspective the length of our lives in comparison to geological time, our fish-like ancestor slowly evolved to live on the land about 375 million years ago. From there, it has evolved into mammals, reptiles, birds, and amphibians. To understand the vast amount of time it has taken for modern humans to evolve, it's helpful to use our road trip example of using distance to understand time. If you were to go on a road trip back in time to witness the first vertebrates on land, using 1 mm = 1 year, you would have to drive 375 km, or about 233 miles to observe our ancient ancestor adapting to live on land. To put that into perspective, the span of our entire life may be less than three inches.

However, just like the theories of Eratosthenes, Copernicus, and Einstein, Darwin's theory of evolution by natural selection has been well supported by more than 150 years of repeated verification from thousands of scientists. Since its publication, evolution by natural selection has become the central paradigm of modern biology and is considered one of the greatest intellectual contributions to modern science. Darwin's theory changed the way we do science by removing the need to use supernatural explanations to explain the natural word.

The Time Was Right for Evolution

Sometimes a scientific theory is proposed at the right time, by the right person, and they receive the majority of the credit for the discovery. Darwin is credited for developing the theory of evolution by natural selection, although he was not the only person to develop this theory. However, he was the right person, in the right place, at the right time to put forth a revolutionary theory that remains with us rather than becoming a side-note in history. In contrast, Eratosthenes was a person ahead of their time and most people have never heard of this Greek philosopher, even though he was the first person to demonstrate the Earth is round. It's just that his work was 1500 years too early!

Sometimes, it can take a long time to change long-held views of society. Prior to Darwin, most people generally assumed the world was young and all the species were created perfectly in the last few thousand years. Many naturalists before Darwin's time, including Linnaeus, wrongly believed that each species of plant or animal was created perfectly in the recent past, and any deviation would be a deviation from perfection. A consequence of this dogmatic view of the natural world was that it slowed scientific advancement for centuries.

Aristotle and other Greek philosophers had a huge influence on western thought for almost two millennia. Their ideas of perfection, an unchanging world, and a young Earth became entrenched in western culture and religious ideology. Unfortunately, few people seriously challenged the dogmatic view of an unchanging world, which slowed and even prevented scientific progress, even though it was not supported by observations and experiments. For example, it took nearly 1800 centuries for Galileo to conduct a simple experiment to disprove Aristotle's hypothesis that heavier objects fall faster than lighter objects, once again an example of science disproving common sense knowledge of the world. Another wrongly-held belief from Aristotle was that women had fewer teeth than men. It only took one person to count the number of teeth between men and women to disprove this idea.

Even today, Aristotle's influence can be seen. If you've ever watched a nature program only to hear the narrator state that some unique animal is "perfectly adapted", then you've witnessed Aristotle's thinking about perfection in nature. In reality, there are no perfectly adapted organisms because there is always a trade-off. If you can run fast, then you lack power; if you are perfectly camouflaged, then you may not be able to move fast.

Beginning in the early 1800s, cracks to the unchanging world view were beginning to emerge from geology. Two geologists, Hutton and Lyell, proposed the theory of uniformitarianism to explain the formation of modern geological features of the landscape such as mountains or valleys. Uniformitarianism states that the same geological processes operating today, operated the same in the past. For example, if the forces that build mountains causes them to grow about an inch per year, then most mountain ranges will grow at approximately the same rate. Or, if erosion carries away sediment from the land, it will do so at similar rates in the past. The implication of uniformitarianism is that it predicts geological processes are slow compared to the life of a human; it takes millions of years to build a mountain or to erode one down. Based on uniformitarianism, Lyell and Hutton were among the first to believe the world was ancient, perhaps tens to hundreds of millions of years old, rather than thousands of years old.

An ancient Earth contradicted many beliefs in the 1800s, including the young Earth where people though the Earth was few thousands of years old. At the time, this was not an absurd idea because no one really had any good way of dating the Earth, so there was no reason to believe the Earth was ancient. It took these early geologists to realize that geological forces were slow, but over time, they could account for fossils of marine organisms on high mountains. By the time Darwin worked on his theory, the age of the Earth was predicted to be approximately 10-100 million years old.

Finally, in the late 1890s the discovery of radioactivity gave scientists a tool capable of determining just how ancient the Earth really is. Although, it took another 60 years to fully understand radioactive decay and half-lives of elements, but eventually the Earth was determined to be approximately 4.6 billion years old. Far older than anyone had predicted or thought possible.

Coinciding with the advent of uniformitarianism in geology, the first scientific hypotheses to explain the origin of species, or how species change over time, were being proposed. They were born out of observations by naturalists and museum collectors who were combing the world, searching for strange new animals and fossils to be cataloged for museum collections. In the early 1800s, the French naturalist, Lamarck published the first scientifically coherent theories of evolutionary theory by stating that the environment causes changes in animals. However, his mechanism of change, acquired characteristics, was later disproved by observations and experiments. Despite proposing an incorrect mechanism for evolution, Lamarck was the first person to use biology as we do today. Ironically, Erasmus Darwin, Charles Darwin's grandfather, also proposed the idea that species could change over time.

Adding to the progress made by naturalist and geologists of time, the economist Robert Thomas Malthus, published a series of articles on population growth between 1798 and 1826. He showed how a population could grow exponentially if there were no deaths from disease, famine, or other causes. He believed that unchecked population growth could lead to catastrophe. Malthus's work would later become influential in Darwin's formulation of evolution by natural selection where he clearly made the point that in nature, more individuals are born than can survive.

By the 1850s, the young Earth idea was being routinely challenged by geologists as it became clear that geological processes are slow. Additionally, naturalists were collecting and studying thousands of plant and animal fossils from around the world, bringing them back to be described and cataloged. Coinciding with advances in natural history, the works of Malthus and Lamarck were being debated in scientific circles. It appeared that the time was right for the right person to put all these pieces together into one revolutionary theory.

Darwin's Journey to Discovery

Charles Darwin was born on 12 February 1809 into a well-to-do family in England, the same day Abraham Lincoln was born. What a great day for the world! Darwin originally started his education by attending medical school, but he quickly lost interest after watching surgery on a young child without anesthesia. Rather, he showed an early propensity to collect plants and animals, similar to the naturalists of the time. In the summer of 1831, one of Darwin's professors presented him with an opportunity to be a gentleman's naturalist and proper companion to Captain Fitzroy of the HMS Beagle. The purpose of the journey was to map the South American coast over a two-year period. At first, Darwin's father was not supportive, calling it a waste of time. Luckily, Darwin's uncle convinced the elder Darwin to let his son go and even to financially support the endeavor.

The HMS Beagle left port for South America on 27 December 1831. While on the journey, Darwin was given a copy of Lyell's *Principles of Geology*, which ended up having a big influence on Darwin's views on the age of the Earth. In fact, on the Beagle's first stop at the Cape Verde Islands, Darwin noticed sea shells in rocks several hundred feet above the surrounding Atlantic Ocean. This was one of the first signs of the Earth's ancient age to the young Darwin, providing more evidence for Lyell's theories.

Unfortunately, Darwin often became sea sick, but when the seas were calm, he spent time collecting plankton, dissecting marine invertebrates, and preparing specimens to be shipped back to museums. He also wrote detailed field notes his observations and findings, which were important in his later writings of evolution after his journey.

Once the Beagle reached the shores of South America, Darwin spent most of his time on land collecting plants, animals, fossils, and minerals. While Darwin made his collections, the Beagle carried out its primary mission, surveying the South American coastline. At times, Darwin would spend several months exploring South America before being picked up by the Beagle to sail further down the coast. During one of his month-long treks near Patagonia in present day Argentina, Darwin discovered the fossils of large extinct mammals; providing another piece of evidence indicating that species arise, change over time, and go extinct.

Some of the most important clues supporting an ancient Earth were found when Darwin was exploring modern day Chile. Along the western edge of South America, the Andes Mountains rise thousands of feet, forced up by the collision of the South American plate into the Pacific Plate. Hiking up into the Andes, Darwin saw fossils of sea shells near the tops of these mountains, far higher than the sea shells he collected on the Cape Verde Islands. While in port, he noticed that a mussel bed had been recently lifted above sea level after a recent earthquake. He realized that given enough earthquakes over a long period of time, mountains could rise carrying sea shells once at the bottom of the ocean to heights thousands of feet above sea level. Observation like this led him to believe that the Earth was old enough for species to slowly change over vast amounts of time.

Perhaps the most famous stopping point of the Beagle's journey were the Galapagos Islands located about 600 miles off the coast of Ecuador. These geologically young islands, less than 5 million years old, have unique plants and animals on each of the main islands. Darwin saw that the plants and animals were unique species to the Galapagos; however, they closely resembled similar species in South America. For example, there were flightless cormorants, giant tortoises, and marine iguanas all similar to species found on the mainland 600 miles away.

HMS Beagle in the Straights of Magellan, South America

As the Beagle sailed across the Pacific Ocean, they stopped at other islands where Darwin made additional collections of plants and animals. Over time, a pattern began to emerge, all the oceanic islands contained unique species; however, island species most closely resembled species to the nearest mainland. Taken together with the evidence of an ancient Earth, these observations inspired Darwin's view that perhaps species could change over time.

The Beagle continued to Cape Town in South Africa and then back across the Atlantic to finish a few surveys of the South American coast before returning home to England. What was to be a two-year mapping expedition of South America, turned into a five-year journey that circumnavigated the world. Finally, the Beagle returned to the same port in October 1836. Upon arrival, Darwin had already gained fame as a naturalist based on his collections he had sent back during his five-year journey. He would spend the next few years writing books about his adventures and the natural history of his collections. In the ensuing decades, he would use his collections, additional observations of selective breeding in cattle, pigeons, and crops, along with the influence of other geologists, naturalists, and the works of Malthus to develop his theory of evolution by natural selection. It took him several decades to accumulate enough evidence to form a well-supported theory that explained his observations, including those of the unique island species that were similar to ones on the mainland.

Darwin's Theory of Evolution by Natural Selection

By the 1850s, it was the right time and Darwin was the right person to bring forth the theory of evolution by natural selection explaining that all modern species descended from common ancestors. Although, Darwin may have been delaying publication or even considered publishing his theories posthumously because he knew that it was going to incur rancorous debates among the scientific elite and would be ridiculed by the public. Fortunately, Darwin's close friends, including Charles Lyell, had encouraged him to publish his work or risk getting "scooped".

In fact, Darwin was not the only one thinking about evolution. In 1858, a young naturalist named Alfred Russel Wallace sent Darwin a manuscript detailing essentially the same theory of evolution by natural selection that Darwin had been working on for decades. Darwin had Wallace's work presented alongside his at the Linnean Society that year. In 1859, Darwin published his book, *On the Origin of Species*, which quickly sold out. Darwin is generally given credit for evolution by natural selection because he had been working on the theory for decades and presented decades worth of support for it. Additionally, there was a clear paper trail showing that Darwin had developed the theory prior to Wallace.

While many naturalists immediately accepted Darwin's work, it still took decades to become fully accepted, partly because no one understood inheritance. Eventually, advances in our understanding of genetics solidified evolutionary theory. In the end, it revolutionized the way we do science and became the central paradigm of the entire field of biology. Darwin's theory that species changing over time, slowly evolving from ancestors, simultaneously explained the unity and diversity of life that we observe today. The basic tenants of the theory have been routinely verified by repeatable observations and experimental data for over 150 years.

Often there is confusion as to whether evolution is a fact or a theory. If we define evolution as species change over time, then it is a fact verifiable by observations and experiments. We can see it happen and measure the rate species change either in the controlled environment of laboratories or in less predictable natural settings. Natural selection is a theory explaining why species change over time. So, evolution by natural selection is both a fact and a theory.

For 3.8 billion years, life has continually evolved becoming abundant and diverse. You can find it on the land, in the sea, or buried deep in the ground. The variety of life in our world is truly stunning, there are whales measuring 100 feet long, trees that live for thousands of years, lichens eking out a living from minerals on rocks, and bacteria living in water hot enough to boil. Everywhere you look, life is present and incredibly diverse. Yet, with all the diversity, an underlying unity reveals evolutionary relationships as modern species descended from common ancestors.

Evolution explains the unity and diversity of life. Darwin, among others, including the taxonomist, Carolus Linnaeus, observed similarities among species. In fact, Linnaeus developed the modern classification of life based on shared characteristics between different taxonomic groups. For example, all animals with hair and mammary glands are classified as mammals based on those shared features inherited from an ancient mammal with those traits. There are many species of mammals because they diversified over time through evolution adapting to exploit different resources in the environment

The theory of evolution was born out of observations. Darwin, like other naturalists, made several key observations about populations. Similar to others that put forth revolutionary theories, Darwin did not develop his theory of evolution by natural selection in a vacuum, he was influenced by his friends and from reading the works of other scholars. First, he observed variation among individuals within a population. You can easily see this just by looking around a classroom or walking into a crowded area, no two people are the same. Second, he observed that more individuals are born in a generation than can survive, which we know from the work of Malthus. Malthus was an economist who developed the principle that unchecked populations would grow continually at an exponential rate. For example, if a pair of fruit flies reproduced and all their offspring survived, and reproduced, and this continued for each generation, then the world would be covered 3 feet deep in fruit flies in less than four months! His work influenced Darwin, because clearly, we are not drowning in a vast sea of fruit flies.

There's more to science than making observations and testing our assumptions, it also relies on imagination and creativity to make new insights or see things in a different way. For example, Darwin made two observations, there is variation among individuals within a population and more individuals are born than can survive. He then realized that survival and reproduction are not random. Evolution by natural selection predicts that the best fit individuals will most likely survive; but most importantly, they are more likely to reproduce, passing on those favorable traits to the next generation. Over time, species change to become better adapted to their environment as they accumulate favorable traits for that environment. Hence the common phrase, survival of the fittest. Darwin used his creative insight to put together his observations into a single testable theory explaining the unity and diversity of life. It's important to note that natural selection acts upon the individual, but only populations evolve. An individual cannot evolve.

Natural selection is a theory because it provides a mechanism explaining how species change over time. A logical conclusion of this theory is descent with modification; species change over time from ancestral species. Therefore, modern species share traits inherited from a common ancestor. For example, all mammals have hair and mammary glands similar to the first mammals that evolved those traits. Likewise, if you were to find a new animal that had hair and mammary glands, then it would be a mammal. Hence, the theory explained the unity of life through descent from a common ancestor.

Today, about 5,416 known species of mammals inhabit every continent except Antarctica. They are a diverse group including whales, bats, cats, and humans. The diversity lies in the fact that the first mammal appeared about 240 million years ago in the Triassic Period, and mammals have been diversifying into different habitats before the dinosaurs went extinct, a very long time! Once again, evolution explained the diversity of life as they descended with modification from a common ancestor.

The Fossil Record Provides Great Support
The fossil record fully supports evolution providing a clear record of descent with modification and extinct species. Combined with geological support of an ancient Earth, it became clear that species would come into existence and then go extinct or evolve into new species. For example, while in South America, Darwin discovered fossils of large extinct mammals. Based on his findings, Darwin predicted that in time, transitional fossils, or fossils that have characteristics of ancient species and modern species would be found. His prediction proved true when the fossil of Archaeopteryx was discovered in Germany around 1861. Archaeopteryx was a transitional species with traits of both dinosaurs and birds. It had feathers and could fly like a bird, but had teeth and a bony tail like a dinosaur.

Layers of rock are stratified with older layers found below younger layers. Therefore, the fossil record makes for an excellent test for descent with modification. If species change over time, then we should find older species in lower geological layers, and modern species in more recent layers. If there is no evolution, then you should find fossils scattered randomly throughout geological layers. In this scenario of no evolution in the fossil record, you could find the fossil of a rabbit next to a dinosaur. However, in every fossil bed in the world, we find that the fossils of ancient organisms coincide with the geological layer they should be in, fossils are not scattered randomly. Instead, the fossil record shows a clear progression in the evolution of life over millions of years. We never find a fossil of a rabbit next to a dinosaur. Therefore, the fossil record totally supports descent with modification and totally refutes the idea that all life was created at the same time.

We place ages on fossils and geological layers by using radioisotope dating. Remember that every element has isotopes, some of which are unstable and decay into other elements at a very constant rate known as the half-life. While some geological layers may be difficult to date, we can piece together the ages of most geological formations by knowing the ages above and below them. It's like putting together a large puzzle; it can be difficult to visualize where an individual piece may go. However, as you put more pieces together, the big picture becomes clearer making it easier to understand where the individual pieces fit.

Because the geological record is layered with oldest rocks on the bottom and younger ones on top, geologists have pieced together an accurate picture of the geological formations and fossils around the world. This was put to the test in the early 2000s when a team of scientists led by Neil Shubin wished to discover a transitional organism linking the evolution of fish to terrestrial vertebrates. They had fossils of fish from 380 million years ago and potential early amphibians from 370 million years ago. Based on the dates of these fossils, they predicted that the ancestor to terrestrial vertebrates may have appeared about 375 million years ago.

To find such a transitional animal, they turned to geological maps to locate the right geological formations that would potentially have fossils from 375 million years ago. In 2004, they planned their trip to the Arctic Circle where they eventually found the fossil they were looking for. They named it Tiktaalik and it represents a transitional organism with both fish-like and amphibian-like characteristics. Finding Tiktaalik was yet another confirmation, along with thousands of other experimental and observational confirmations of evolution and descent with modification. Neil Shubin went on to write a book called *Your Inner Fish* about finding Tiktaalik and its implications for our understanding of vertebrate evolution.

Phylogenetic Trees Illustrate
Evolutionary Relationships

Perhaps, Darwin's most remarkable insight was that all modern species descended from ancestral species. Over time, species slowly changed due to relentless natural selection. Life diversifies as species evolve, adapting to their environment. Because modern species evolved from past species, they share a common ancestry and therefore share traits in common. The more recent a common ancestor between two species, the more closely related they will be to each other, and the more similar they will appear.

I'll use the three species of rosy-finches in North America to explain evolutionary relationships. These small song birds nest above the tree-line in the mountains of the western North America. The extensive glaciation split the ancestral population of modern Rosy-Finches into at least three different populations. With the populations isolated from each other and no longer able to interbreed, small changes began to accumulate in the populations, slowly making them different from each other. Once the ice retreated, the populations were sufficiently different that they no longer interbred, even if they lived in the same areas again. As a result, there are now three species when historically there was only one.

Based on shared characters of organisms, we can construct phylogenetic trees, which shows our best hypothesis of their evolutionary relationships. Rosy-finches are related to other songbirds, such as the Northern Cardinal, but millions of years of evolution separate a Northern Cardinal from rosy-finches, so they share less in common with each other. We can take this a step further, all birds have feathers and a bill made of keratin that lacks teeth, indicating that one common ancestor about 135 million years ago gave rise to all modern birds. Go back further in time, and birds and crocodiles shared a common ancestor more than 240 million years ago. In that time, birds and crocodilians have greatly diverged, yet they share common characteristics, including a four-chambered heart and other behaviors including making nest for their young, parental care, and they sing to their mates. Although, it's hard to call an alligator's deep guttural vocalizations a song.

Alligators and birds are living descendants of a much larger and once diverse group of reptiles called the Archosaurs (arch-ruler, saur – lizard, or ruling lizards), which included the dinosaurs. Dinosaurs went extinct about 65.5 million years, making it hard to find clues to how they lived. Phylogenetic trees help us make predictions about the behavior and characteristics of extinct animals by comparing groups of living animals. In our bird and crocodile example, we know they share several behaviors, including making nest, parental care, and singing to their mates. Because dinosaurs were Archosaurs, we can predict that they most likely made nest, had parental care, and possibly sang to their mates. Two of these predictions have been verified with the finding of fossilized dinosaur nests with females still on top of their eggs. Birds and crocodiles both have four-chambered hearts, so we would hypothesize that dinosaurs also possessed four-chambered hearts, lending credence to modern theories that dinosaurs were active animals.

If you go back to the Pennsylvanian (part of the Carboniferous Period) some 320 million years ago, you would find the ancestor to both reptiles and mammals. It reproduced internally and protected its embryo in an amniotic sac. Amniotic sacs provide a watery environment to protect the growing embryo and along with internal reproduction allowed these animals to survive in a drier environment by reducing their dependence on returning to water for reproduction. Go back another 50 million years, and you would find the ancestor to all tetrapods (tetra = 4, pod = foot) living along stream banks in vast swampy forests of the Devonian around 370 million years ago.

To this day, the tetrapods include amphibians, reptiles, birds, and mammals, most of which possess four limbs (or have lost them as in snakes and whales). Not only do they possess four limbs, they also have the common bone pattern of one bone, two bones, lots of bones;

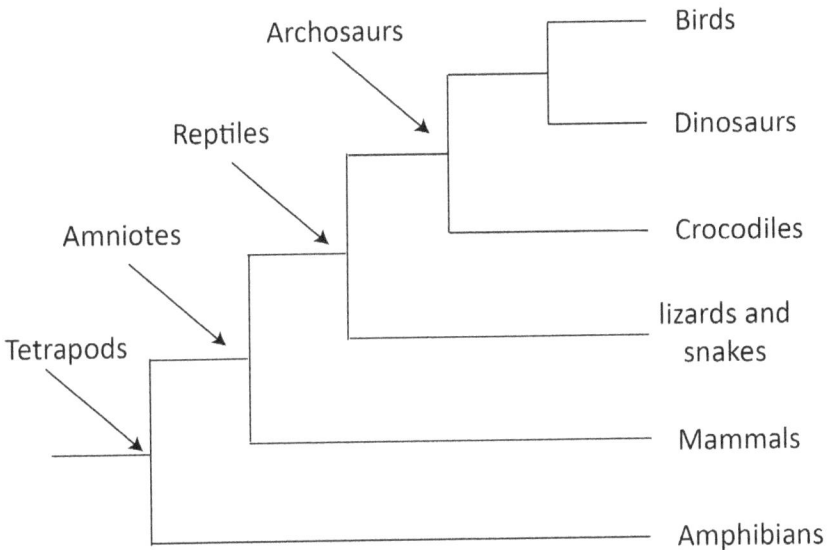

A phylogeny representing the relationship among terrestrial vertebrates. The arrows point to a common ancestor for each lineage. For example, mammals and reptiles both amniotes, they have an amniotic sac and reproduce internally.

a feature we inherited from that distant ancestor with both fish and tetrapod characteristics.

Continuing our journey back in time to 500 million years ago to the Cambrian, you would find small "worm-like" animals that were the ancestor to vertebrates, which include all sharks, fish, amphibians, reptiles, birds and mammals. Similar to us, they had a dorsal nerve cord, a notochord, muscular tails, and pharyngeal slits modified into gills for breathing. These defining characteristics of vertebrates set us apart from all other animals. In the tetrapods, the gill slits were modified to form parts of the neck, jaws, and inner ear bones.

In the last 500 million years, those features of the first vertebrates became greatly modified to accommodate different life styles. If you go back even further, to 550 million years ago near the end of the Proterozoic, you would find primitive animals with true tissues, muscles, nervous system, and they were eating by ingestion. Today, all animals from spiders, to snails, to earth worms, to birds and mammals, all possess traits that were present in our most ancient animal ancestors. By comparing the similarities of all living animals, we can make certain predictions about our common ancestor. So far, all the evidence we have collected, including the fossil record, shared morphological traits, and the history written in our DNA, all tell the same evolutionary story. No other theory, idea, or ideology makes such powerful predictions of evolution that have been supported by repeated observations and tests for over 150 years.

Scientific Challenges to Darwin's Theory

Despite the predictive power of Darwin's work and the enormous amount of support he presented for the theory, there were several valid criticisms at the time of his publication in the mid-1800s. Most of the scientific criticisms stemmed from a lack of knowledge about genetics and inheritance. For example, no one knew how traits were passed from one generation to the next. At the time, most people thought traits were blended, known as blending inheritance. Critics of natural selection argued that any variation would be swamped out over generations, so any positive change would not accumulate in populations. Additionally, no one knew about mutations creating variation in a population. Without a continual source of variation, evolution would eventually grind to a halt because there would be no variation to act on.

Some critics also thought the Earth was much too young for all the species present today to have evolved slowly over time. In fact, the idea of a young Earth is still pervasive in society today despite enormous amounts of data to utterly refute it. But, in the 1860's there really was no way for geologists to accurately age the Earth, it took the discovery of radioactivity, the calculations of half-lives, and nearly 100 years of work to accurately determine the age of the Earth. Much to the surprise of geologists and physicist working on the problem, multiple lines of evidence support the Earth to be 4.6 billion years old. An age that was far older than anyone thought at the time.

Even today, critics point to the incompleteness of the fossil record wrongly believing it lacks transitional species. While it is true, we lack a fossil record of every species that has ever existed on the Earth. In fact, it is biased towards animals with hard parts that easily fossilize. Despite these shortcomings, the fossil record fully supports evolution as explained in the previous section. By finding older fossils below younger fossils, we see a clear progression in age based on their position in geological layers. If you were to hike down the Grand Canyon, you would be hiking back in time, continually finding older fossils as you hike deeper into the canyon. Fossils are not randomly distributed in the geological strata.

Lastly, many people just simply did not accept evolution because it was contrary to their everyday experiences, it went against their common-sense knowledge. However, our lifespan is only a tiny fraction compared to geological time scales that span millions, or even billions of years of evolutionary history. Additionally, the *zeitgeist* of western civilization was influenced by Aristotle who believed that all species were created perfect and that any deviation would be a deviation from perfection. As a result, many rejected the idea that species change over time because they believed the world to be relatively young and unchanging.

In the last century, the weight of scientific evidence has completely disproved these criticisms. Evolution is well established and we know that species change over time is an indisputable fact. We can directly observe evolution occurring naturally or in the laboratory; we can measure the rate of evolutionary change in organisms; and we can test for it to experimentally verify it. Darwin's theory of natural selection, which explains how evolution happens, has been repeatedly verified by thousands of independent observations and experiments and represents a consilience of knowledge. Today, evolution by natural selection is one of the most well supported theories in science, along with relativity, quantum mechanics, and the Big Bang.

Evolution by Natural Selection
The Theory that Revolutionized the World

Darwin's theory of evolution by natural selection caused a paradigm shift in the way we do science. Although Darwin wasn't the first naturalist to discover evolution by natural selection, in the 1840s, a forester named Mathews devised practically the same theory as Darwin, but thought that evolution by natural selection was such common sense, he didn't publish it. Despite not publishing his theory, he thought he should have received credit for it. More than a decade later in the 1850s, Alfred Russel Wallace came up with evolution by natural selection and sent his theory to Darwin. Luckily for Darwin, he had a long, written history of developing his theory first, so he gets the majority of credit for evolution.

Evolution by natural selection was important for two reasons. First, it was an elegant theory that explained the unity and diversity of life. It fit with the Linnaean classification of life and explained the patterns in the fossil record. It also has important applications including cancer research, the emergence of antibiotic resistant diseases, conservation, and agricultural practices.

The second major contribution was that it fundamentally changed the philosophy of science by showing that scientific explanations can be used to explain all natural phenomena. Prior to Darwin's time, most naturalist, including Linnaeus, believed that all life was created basically in its present form at some fixed time in the past, a belief that was a hold-over from Aristotle. No one really searched for answers as to why there are so many species or for the origins of life. If they did, supernatural explanations outside of the realm of science were often used. By developing the theory of evolution by natural selection, Darwin demonstrated we did not need supernatural explanations to explain the natural world. Rather, the answers could be determined based on scientific observations that are measurable and repeatable. Using a scientific explanation to explain the origin of species resulted in a profound paradigm shift that spread to all branches of science. It was an intellectual revolution paving the way for the most rapid expansion of scientific understanding in the history of our species.

Charles Darwin Circa 1854 when he was preparing
On the Origin of Species

Chapter 5

The History of Life is Written in Our Genes

There are more than 3.8 billion years of evolutionary history written in our genes

Introduction

The history of life is written in our genes, contained in the sequence of just four letters known as nucleotides. With today's technology, we can rapidly sequence entire genomes, shedding light on evolutionary history and strongly corroborating the fossil record in support of evolution. Modern sequencing has allowed us to determine the relationships among modern organisms and even provide rough estimates of when they last shared a common ancestor with related species.

By comparing genomes of many species, we discovered that every organism on this planet including plants, fungus, and even simple bacteria shares some genes in common. These shared genes have been passed down from generation to generation going back billions of years. We also possess the remnants of genes that no longer work, but were once used by our ancestors millions of years ago. Additionally, the genes of ancient viruses continue to reside in our genome, and we do not know exactly why.

If only Darwin knew about modern genetics when he wrote *On the Origin of Species*, it would have added even more support to his theory and instantly silenced many critics. At the time of his writing in the 1850s, blending inheritance was the prevailing hypothesis of heredity, so any beneficial variation would be swamped out within a few generations. Adding to the problem was that no one knew about mutations creating variations in a population; as a result, Darwin's theory was vigorously debated for decades. However, by the 1940s, 80 years after Darwin published his theory, the Modern Synthesis emerged when Mendelian inheritance, modern genetics, and Darwin's theory of evolution were combined into one comprehensive theory largely settling the evolution debate. Although, work on evolution did not stop, in some ways, it even sped up, scientists just weren't trying to prove it happened.

By the 1950s, it was known that DNA, held the genetic information necessary to make every living organism. The discovery of DNA's structure and the genetic code a decade later have been considered among the greatest scientific advancements of the 20th century. By making these discoveries, it linked heredity and evolution directly to molecular biology.

Today, we are in the midst of another revolution in science as our understanding of how genomes work continues to grow at a fast pace. We can rapidly and cheaply sequence DNA, and the cost of genetic sequencing continues to decline. Already, some doctors sequence the genomes of patients to determine what diseases they may be susceptible to or which cancer treatments may work best. With the advent of modern genetics, we can now create genetically modified organisms by inserting genes of one organism into other organisms. Recent technological breakthroughs using CRISPR technology allow us to edit genomes with incredible precision, which is already creating major advancements in medicine and the development of genetically modified organisms (GMOs). To begin our understanding of heredity, I start with the pioneering works of Gregor Mendel and Thomas Hunt Morgan.

Heredity

For more than 3.8 billion years, the continuity of life has remained intact as the information in DNA has been faithfully replicated from one generation to the next. The implications of this statement is that you represent an unbroken lineage going back more than 3.8 billion years. The passing of genetic information from parents to offspring forms the basis of heredity. However, we haven't always known how heredity works. When Darwin published his theory of evolution by natural selection, there were several major gaps in knowledge, reflective of the times. First, no one knew how traits were passed from one generation to the next. Second, no one knew how variation in a population could arise. If you think about humans, what's the source of black hair, blond hair, or red hair? Fortunately, both of these questions have been answered with advancements in our understanding of genetics.

In the 1860s, debates on evolution by natural selection focused on how new traits could arise and be passed to the next generation. It would take the pioneering work of Gregor Mendel and Thomas H. Morgan to shed light on the nature of inheritance and how it supports evolution. By experimenting on plants, Gregor Mendel, disproved blending inheritance and showed that traits are passed from one generation to the next as discrete units. At the start of the 20th century, Thomas H.

Morgan discovered mutations as the source of variation in populations and showed that genes are located on chromosomes. By the 1940s, the theory of evolution by natural selection was combined with the work of Mendel, Morgan, and others into a "Modern Synthesis", a term coined by the biologist Julian Huxley. You may recognize the name Huxley, Aldous Huxley was his brother who wrote the book, *Brave New World* published in 1932.

Gregor Mendel – 1860s

The age of modern genetics began in the 1860s with Gregor Mendel, but his work remained mostly unknown for 30 years until its rediscovery in the 1890s. Mendel experimented with pea plants to determine how traits are inherited (today we refer to traits as genes and different versions of trait are alleles, I will use the modern terminology from here). In mid 1800s, most generally assumed that genes were blended from one generation to the next, like mixing colors in paint. If there was variation, then it would be diluted over several generations, eventually being lost. This was one of the main criticisms of Darwin's theory. Mendel knew that organisms had genes that could be inherited, for example pea flowers were either purple or white. However sometimes, the white allele would seem to skip a generation. This observation didn't follow the predictions of blending inheritance.

To understand how genes are inherited, Mendel performed numerous breeding experiments crossing pea plants following the inheritance of different genes (flower color, seed color, seed shape, *etc.*) In one experiment, Mendel crossed a true breeding purple flower with a true breeding white flower and found that all the offspring had purple flowers. Rather than stopping the experiment, Mendel continued by crossing the first generation (F1) of all purple flowers. In the second generation (F2), both versions of the alleles were present; approximately 75% of the plants had purple flowers and approximately 25% had white flowers. The white flower trait, apparently "skipped" a generation. Fortunately, Mendel was a good at math and saw the significance of the 3:1 ratio of purple to white flowers in the second generation.

Mendel realized that each parent contributed one allele to the next generation for any gene. Meaning there is a gene for flower color and there are two versions of the gene (purple and white), or two alleles. To explain how the white allele skipped a generation, he concluded that some alleles were dominant, and others were recessive. In this example, the purple flower color was dominant, requiring only one copy to make a purple flower, whereas the allele for white flowers was recessive. Therefore, to have white flowers, a plant needed two copies of the white allele. That's why the first generation of pea plants in his experiment produced all purple flowers because each plant had one copy of the purple allele. When breeding the first generation with themselves, about 25% of the offspring received two copies of the white allele and made white flowers.

Today, Mendelian inheritance refers to a specific scenario where one gene determines one feature, and each gene has two alleles. The alleles can be either dominant or recessive. In the pea flowers, there are two versions of the gene for flower color, a white allele and a purple allele. Also, alleles are separated independently from each other during gamete formation and genes are also inherited independently of each other. For example, flower color has no influence on the inheritance of seed color or seed shape. However, I should point out that many genes do not follow Mendelian inheritance.

The results of Mendel's experiments can be applied to other sexually reproducing organisms including humans. In sexual reproduction, males and females produce gametes, each carrying one copy of their parent's genome; the gametes combine to form a new offspring. Therefore, sexually reproducing organisms carry two copies of their genome, one from each parent, meaning we are diploid. Humans have 46 chromosomes, or 23 pairs of homologous chromosomes. Homologous chromosomes are the same length and have the same genes in the same location on the chromosome. Because we have two copies of our genome, we always have two alleles, they can be the same, or two different alleles.

If an organism has two different alleles for a gene, where one is dominant and the other is recessive, then organism won't show any sign of the recessive allele. This was the case for the first generation of pea plants that were all purple; the offspring had one allele for white flowers, and one allele for purple flowers. The second generation (F2) would have individuals with three combinations of alleles (two purple, one purple and one white, and two white) resulting in plants with purple and white flowers.

Perhaps the most important finding of Mendel's experiments was the discovery that genes are discrete entities, passed in whole from one generation to the next, clearly supporting particulate inheritance. Mendel's findings disproved blending inheritance that plagued Darwin's theory of evolution by natural selection. Because genes are discrete units, they won't necessarily get "swamped out" in a population. In fact, favorable genes are more likely to persist and increase in frequency in a population over time. Whereas bad genes will tend to decline or even perish over time. Hence, species change as they accumulate favorable genes improving their "fit" to the environment.

To fully understand the heredity of flower color in Mendel's pea plants, we can combine genetics with chemistry, knowledge that wasn't present in Mendel's time. Pea plants with the purple allele produce a pigment called anthocyanin that builds up in the flower pedals turning them purple. Anthocyanin is the same pigment that gives blueberries and other fruits their dark blue color. At some point in the past, a mutation occurred in a pea plant, damaging a gene used to make anthocyanin, resulting in a new allele that makes white flowers.

In pea plants, they only require one working copy of the gene to make anthocyanin, resulting in purple flowers. Because only one copy is required for the gene to be expressed, it is considered a dominant allele. To have a white flower, the plant needs two copies of the white allele, so it is considered a recessive allele. It has nothing to do with one allele being more common or stronger than another allele.

The mutation causing the white allele added genetic variation to the pea plants and was not removed from the population because insect pollinators also come to white flowers. Keep in mind that Mendel really didn't understand what a gene was or the source of alleles in a population, and neither did Darwin. That understanding would begin to emerge 40-50 years later in the early 1900s with the pioneering work of Thomas H. Morgan and his gifted graduate students, some of who went on to win Nobel Prizes.

Mendel's experiments with pea plants provided the first rigorous test of inheritance. But, to be blunt, he also got lucky by picking genes that followed very simple rules of inheritance. This pattern does work for a very small number of genes. However, by the turn of the century, work by Morgan showed that patterns of inheritance are much more complicated than depicted by Mendel's experiment. Even though Mendel's view of inheritance was very simplistic and far from the complete picture, he was able to use his results to disprove blending inheritance by showing that a gene is inherited as a discrete unit. His contributions to our understanding of heredity fully supported Darwin's theory of evolution by natural selection.

Thomas Hunt Morgan and the Fly Lab - 1910s

While Mendel knew that traits were passed from parents to offspring, he didn't understand how or where they were located. Part of the answer began to emerge in the 1870s with the discovery of chromosomes and two types of cell division in eukaryotes, mitosis and meiosis. Mitosis produces two genetically identical offspring. Prior to mitosis, the DNA is replicated and then condenses into visible chromosomes. During mitosis the chromosomes separate forming two genetically identical daughter cells. Almost every single cell in your body reproduces by mitosis. Because there is no recombining of genomes, and both daughter cells are mostly genetically identical.

Meiosis was discovered in cells that form gametes, the eggs and sperm. It takes two rounds of cell division and results in 4 daughter cells that are genetically different. Each gamete carries only one copy of the genome, instead of two copies. The discovery of gametes, carrying only a single copy of the genome supported Mendel's hypothesis that sexually reproducing organisms receive a copy of each gene from one parent. Further experiments determined that chromosomes were made of proteins and DNA. Although, at the time, no one was sure how genetic information was stored, or even which of the two molecules was responsible for housing the genetic information.

The next major step in our understanding of genetics began in the early 1900s from studying fruit flies (*Drosophila melanogaster*). It's amazing that much of what we know about genetics has come from studying a small fly that thrives on ripe fruit. Fruit flies are what we call a model organism and have been a favorite of genetic studies for more than a century. They are a model organism because they are easy to keep and rapidly breed in captivity. Second, they have 4 pairs (8 total) of large chromosomes, making them relatively easy to work with.

The importance of fruit flies to our understanding of genetics began with a single observation of eye color. The vast majority of fruit flies have red eyes, but one day, Morgan noticed an individual with white eyes. He began to wonder what caused the rare white-eye condition and whether or not it was heritable. To answer his question, he began breeding fruit flies by the tens of thousands in his lab, eventually a few mutants with white eyes randomly appeared seemingly out of nowhere. The discovery of white-eyed mutants confirmed that mutations appear randomly providing a source of variation in populations. After additional breeding experiments with the white-eyed mutants, it became clear that chromosomes contained genetic information and the pattern of inheritance for eye color didn't follow the simple rules of Mendelian inheritance, rather eye color was linked to the sex of the fly.

In addition to eye color, Morgan's work on fruit flies showed that many genes are inherited together because they are located on the same chromosome. In Mendel's pea experiment, the genes for seed color, seed shape, and flower color are all inherited separately because those genes are located on separate chromosomes. But, this is not the norm for most genes. Think about it this way; you have 46 total chromosomes, 23 pairs of chromosomes. On those 23 pairs of chromosomes there are about 20,000 genes.

The implication here is that a single chromosome holds hundreds to thousands of genes, which are inherited together, most of the time. Using humans as an example to show how genes are inherited together, the gene for seeing red and green colors is located on the X chromosome. Therefore, color blindness is linked with gender because the gene is physically located on X chromosome that helps determine sex. If you're a color-blind male, then you inherited color blindness from your mom.

By the 1920s, Morgan's group had greatly expanded on the early work of Mendel. They discovered mutations are the source of variability in a population by producing new alleles, the chromosomal basis of inheritance (genes are located on chromosomes), that many genes are inherited together, and his student Hermann Muller created the first maps of genes on chromosomes. The work of Morgan and his graduate students was instrumental in bringing in the modern age of genetics and reconciling the scientific criticisms of natural selection with a modern theory of evolution. Ironically, early in his career, Morgan was a harsh critic of Darwin's theory of evolution. However, after his pioneering work on fruit flies, he changed his mind once he realized the connection between heredity, genetics, and evolution.

Over time, our knowledge of genetics and heredity has continued to grow adding more extensions to Mendel's work. For example, there can be more than two alleles for a gene. Keep in mind than an individual can only have two alleles, but there can be many alleles for a gene in a population. We can use the 'ABO' blood type as an example. Imagine the scenario where your mom has type 'o' blood, your dad is type 'AB', and you are 'A' blood. In this case there three possible alleles, but since 'o' blood is recessive, your mom would only donate the allele for 'o' blood. Your dad possesses two alleles, 'A' and 'B', so if you are blood type 'A' then you would have the recessive 'o' allele from your mom and the dominant 'A' allele from your dad. However, you could have also received the 'B' allele from your dad. As you can see, there are more than two alleles possible, 'A', 'B', and 'o'.

The 'ABO' blood type also offers another extension to Mendel's inheritance. Recall that in our imaginary scenario, your dad is blood type 'AB' meaning he is expressing both the 'A' allele and the 'B' allele. In this case, they are codominant meaning they are both expressed in an individual. Other alleles exhibit incomplete dominance because they appear blended. When a red snapdragon is crossed with a white snapdragon, the offspring all have pink flowers. In this case, it takes two functioning alleles to form a red flower. Only one functioning allele produces half the red pigment, so the flower is pink. Sometimes, many genes work together for a particular trait like eye color where up to 15 genes may be involved, explaining why there is so much variation in eye color among humans. In contrast, a single gene may affect numerous traits, like the sex-determining region of the y chromosome (SRY-gene), which turns on other genes early in development and turns a female fetus into a male.

The Importance of Mutations for Evolution

The thought of mutations often conjures images of the super cool Teenage Mutant Ninja Turtles, the X-Men, a three-eyed fish, or glow-in-the-dark animals. Other than Ninja turtles or X-Men, we usually think mutations are bad and rightly so because some mutations, even very small ones, are lethal. However, they are not always bad and contrary to popular belief, mutations are important because they provide the ultimate source of genetic variation within a population. Mutations in genes form new alleles as we saw in the purple and white pea flowers, the red and white eyes of fruit flies, or the ABO blood type in humans. Without variation, populations could not evolve because every organism would be identical and there would be no natural selection driving adaptations. Clearly this is not the case, we see immense diversity in nature. Over billions of years, mutations have provided the variation needed to evolve organisms as different as humans, fungi, plants, and bacteria.

A mutation is any change to the sequence of DNA, ranging from small to the very large. Small mutations include the switching of one nucleotide for another, or the loss or addition of a single nucleotide. Imagine the word 'out', change the 'o' to a 'c' and you have a new word, 'cut'. Although these are small mutations, their effects on an organism could range from beneficial, to nothing, to lethal. Larger mutations also include the duplication of genes, whole chromosomes, or even entire genomes.

In the big picture, mutations can be harmful, neutral, or even beneficial. Sometimes mutations can be harmful in one environment, yet be beneficial in others. One example is skin color in people. The first humans evolved near the equator in Africa where you get 12 hours of sunlight every day, which is quite harsh on our skin. As our ancestors became less hairy, they evolved to produce more melanin to protect their skin from the damaging effects of ultraviolet light. If you've ever gotten a tan, it's because your skin began to produce melanin to protect itself from the sun.

Over thousands of years, humans migrated out of Africa and northward into Europe and Asia where there is much less sunlight, especially in the winter. Very dark skin in these higher latitudes can be bad because it reduces Vitamin D production, which is not good for our health. In some groups of people, the genes responsible for producing melanin in our skin, hair, and eyes, mutated leading to people with light skin, blond hair, and blue eyes. In this case, mutations creating new alleles with less melanin led to pale skin, allowing for more Vitamin D production, were beneficial and kept.

Beneficial alleles accumulated in our ancestors' populations as they adapted to the northern environments. On the other hand, if the same mutation causing pale skin occurred near the equator, it would be harmful and wouldn't last long in the population. As you can see, skin color is merely an adaptation in humans to protect us from sunlight, the mutations for lighter skin color can be beneficial or harmful depending on the environment.

Some mutations at first may be neutral, providing no immediate benefit to an organism so natural selection does act on it. For example, gene duplication creates families of genes increasing the size of genomes and the number of genes. Once a gene is duplicated, it can mutate, forming a new gene. If the new gene becomes beneficial, then it will be kept and spread through the population over many generations. Gene duplication has been an important in the evolution of animal life by providing more genes to work with, allowing animals to evolve to be more complex.

Have you ever noticed that the family dog seems to sniff everything around him, indeed dogs have an excellent sense of smell. The ability of mammals to detect odors is the result of many gene duplications, leading to a large gene family. A good sense of smell depends on the number of odor receptors you have, which of course, is determined by genes. Early in the history of mammals, the genes responsible for smell duplicated over a thousand times. Each time a mutation created a new copy of the odor detecting gene, it was subjected to mutations and natural selection. What entailed was the evolution of a gene family allowing mammals to detect an incredibly wide range of scents.

Humans don't have the same sense of smell as a dog. But, we actually have the same genes or, more accurately, remnants of the genes that dogs have to detect odors! Instead of having nearly a thousand functioning genes, we have about 500 functioning genes used for smell. The rest mutated and no longer function, so they are called pseudogenes.

Part of the answer to why we have lost our ability to detect odors lies in a trade off with improvements to our eyesight. Primates see more color than dogs and other mammals because the genes used to make proteins to detect light and color were duplicated in the ancestor of primates about 60 million years ago. This gene duplication is similar to the gene duplication that led to better sense of smell, except in this case, it led to a better sense of vision. The gene duplication coincided with the evolution of flowering plants and the nutritious fruits they produce.

A gene duplication in the proteins used to detect light allowed primates to see more colors, coinciding with changes in diet, think about all the colors of the fruits and vegetables we eat. Through mutations, our ancestor's eyesight improved while the necessity to detect odors became less important. When a mutation knocked out a gene required to detect a particular odor, it wasn't bad for our ancestor, so natural selection never acted upon it and the broken gene was passed on to the next generation rather than being removed from the population.

Although they are mammals, whales and dolphins can't smell anything in their environment. Their loss of smell provides an extreme example of relaxed natural selection, where mutations resulted in loss of function with little consequence. Because they are mammals, they have all the gene remnants to smell, but over evolutionary time, those genes mutated resulting in pseudogenes that no longer function. But, that's OK, they don't need to smell in their environment to be successful. Natural selection did not remove those mutations from the population because they were of little or no consequence.

The Structure and Replication of DNA - 1950s

By the early 1950s, it was well-known that genes were located on chromosomes and experiments from the 1940s showed that DNA and not the proteins stored genetic information. However, no one knew the structure of DNA or how it stored genetic information. In the early 1950s, the race was on to discover the structure of DNA and how it stores information. Developed in the 1940s, a technique called X-ray crystallography determines the shape of molecules by shooting high energy X-rays at a crystalline form of the molecule under study. Early on, it took a lot of time and effort, sometimes taking several years to obtain one good image of a single molecule. Once a usable image was made, it required expertise in chemistry and physics to correctly determine the shape of the molecule.

Unfortunately, the discovery of DNA is not without its controversy. Rosalind Franklin was an expert at X-ray crystallography, and her research lab at Kings College in London took one of the best X-ray images of DNA. In 1953, the X-ray image known as Photo 51 was given to James Watson and Francis Crick without her knowledge. Using Franklin's image, they determined the structure of DNA and quickly published the results in the prestigious journal *Nature*. Follow-up work for the rest of the 1950s supported their findings, firmly establishing the importance of DNA as the molecule of heredity. Less than a decade later, in 1962, Crick, Watson, and Wilkens each shared in the Nobel Prize for "their" discovery of the structure of DNA. Unfortunately, Rosalind Franklin was not awarded the Nobel Prize, she had passed away four years earlier from ovarian cancer.

DNA Structure

Discovering the structure of DNA has been hailed as one of the most important discoveries of the 20th century. Its shape, a double helix has been practically synonymized with scientific advancement. DNA is a very large molecule made of just four building blocks called nucleotides (adenine, thymine, guanine, and cytosine). The number of nucleotides determines the length of DNA, which can be highly variable among species. An average bacteria, has about 5 million base pairs, but the length can be over a thousand times longer in plants and animals. In humans, our DNA includes about 3 billion base pairs if stretched out, it would about 6 feet long.

DNA is a double-stranded molecule held together by weak chemical bonds between the four nucleotides. To visualize the structure of DNA, imagine taking a ladder and twisting it to form a double-helix. The inside "rungs" are formed by weak chemical bonds between the nucleotides of the two strands. In any strand of DNA, the base pairing between the nucleotides is always the same: thymine always forms two hydrogen bonds with adenine (A=T) and guanine always forms three hydrogen bonds with cytosine (G≡C).

As a result, both strands contain the same genetic information. The weak bonds between the two strands are easily broken and reformed, an important aspect of the structure allowing the two strands of DNA to be easily replicated or transcribed, a process I will discuss next. The outside part of the double helix, holding the "rungs" are formed by a sugar-phosphate backbone held together by stronger chemical bonds (covalent bonds for those interested in chemistry) that do not easily break.

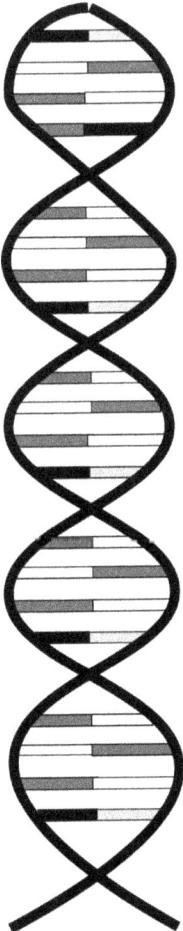

DNA Replication

If there is to be continuity of life, then the information contained in DNA must be accurately replicated to form the next generation. Once they discovered the structure of DNA, Watson and Crick quickly deduced how DNA was replicated, which was verified by experimental data. DNA replication requires the coordinated effort of at least nine proteins in addition to environmental signals triggering cells to replicate their DNA.

DNA replication begins when proteins separate the two strands of DNA, unwinding the double helix, allowing another protein called DNA polymerase to make a new daughter strand of DNA from each of the parent strands. The copying of DNA is very precise, with only about one mistake in 100,000 base-pairs. Replication

DNA
Nucleotides
Guanine
Adenine
Cytosine
Thymine

makes two new daughter strands of DNA. Proteins proof read the DNA, making sure that the information is correctly copied.

The DNA in our cells is chopped up into multiple linear chromosomes where the ends of the DNA are free. The linear structure of our DNA has a major implication: DNA polymerase, the protein responsible for copying DNA, cannot fully copy a linear strand of DNA to the very end. Every time DNA is replicated in our cells, a small portion of the ends are lost, shortening the DNA with each generation. This means our cells have a limited number of cell divisions.

Eventually, after many rounds of replication, the DNA will shorten to the point that it damages the genes required for the cell to survive, causing it to eventually die. To prevent a loss of information from the shortening of DNA, eukaryotes use a repeating unit of nucleotides called telomeres at the ends of their chromosomes. Each time DNA is replicated, the telomeres become shorter resulting in one less replication our cells can undergo.

If there is a rule of nature, it would be that there are no perfectly adapted organisms, every innovation comes at cost. The same goes for multicellular organisms including animals. One cost of being a multicellular organism is cancer, the uncontrollable growth of cells that if left unchecked causes death to the animal. We have more than 10 trillion cells in our body whose replication must be precisely regulated. If just one of our cells becomes damaged and escapes our body's controls on replication, it can become cancerous. With the discovery of shorter telomeres in older eukaryotic cells, it has been hypothesized that linear DNA in eukaryotes is an adaptation for animals by limiting the number times our cells can replicate to prevent cancer. If cells have a limited number of replications, then cancerous cells would eventually die after a few rounds of replication. The trade-off here is that linear DNA places an upper age limit on animals. The cells in your body only have so many replications before they will die

But, we are not born old, somehow the biological clock gets reset with every generation. One way to reset the clock takes place during gamete formation, animals produce a protein called telomerase that lengthens the telomeres at the ends of the cell's DNA. By the time you are born, the genes that make telomerase are permanently turned off in all your cells.

Unfortunately, mutations in some cancerous cells turn on the gene to make telomerase, allowing cancer cells to replicate forever by continually replacing their telomeres. As you can imagine, studying the regulation of telomerase is an active area of cancer and aging research. After all, wouldn't it be great to reset our cells and remain young for a number of years, or be able to stop cancer growth?

Gene Expression
From Genes to Proteins - 1960s

After determining the structure of DNA in the 1950s, the next big discovery was to determine how DNA stored information and how that information makes proteins. Gene expression is the flow of information from our DNA to proteins. At its most basic level, information stored in DNA is transcribed to an intermediate called mRNA, and then the information is translated into proteins in a large molecular complex called a ribosome. Cells can fine tune both transcription and translation to rapidly respond to changes in their environment, or in multicellular organisms they can express unique combinations of genes so that cells can form different tissues such as skin or muscle. A gene has traditionally been defined as a sequence of DNA that codes for a product such as a protein. However, this simplistic notion has been challenged as we have learned that a large portion of our genomes code for short sequences of RNA involved with gene regulation and are not coding for proteins.

The Genetic Code

Sometimes, the beauty of life lies in its simplicity. Recall that DNA stores genetic information in the sequence of just four different nucleotides: adenine (A), guanine (G), cytosine (C), and thymine (T). Take a moment to reflect on the implications of this statement: All the information required to make every protein, in every living organism, in millions of species from bacteria to humans, is stored in the sequence of just four letters. The English language with its tens of thousands of words, and almost limitless stories uses 26 letters, but DNA, the universal language of life only uses 4 letters to make every protein and organism on the planet!

The code of life was cracked in the 1960s when scientists determined how the four letters of life, A T G C, form words known as codons. Codons are a sequence of 3 nucleotides that codes for a specific amino acid (recall that amino acids are the building blocks of proteins). There are 64 possible codons (4 x 4 x 4 = 64), but 3 are stop codons, which tells a ribosome to stop making a protein. Recall that about 20 amino acids used by all life, and an average of three codons for each amino acid, making the genetic code redundant. Additionally, the same codon always codes for the same amino acid, so the genetic code is also unambiguous.

For the most part, all life uses the same 20 amino acids, meaning that when you look at the trees outside, they use the same genetic code as you, and they use the same 20 amino acids to make all their proteins! Although, every protein is made using the same 20 amino acids, millions of proteins can be formed by varying the sequence of amino acids determined by a single, universal genetic code conserved across all life.

The universality of the genetic code among all living organisms is another piece of evidence supporting the theory that all life descended from a common ancestor. If only Darwin had such information. It would have quickly ended many rancorous scientific debates that stemmed from a lack of genetic knowledge. Additionally, there are several medical and technological implications of a universal genetic code that I will discuss at the end of this chapter.

Many people think of the genetic code as the blue print of life. However, this is not a good analogy; a blueprint is a precise set of directions to build some object such as a house or a cell phone. A much better analogy is to think of the genetic code as your grandmother's cook book containing recipes that work together to create an awesomely huge Thanksgiving dinner. And if you've ever used a cook book, you may have made small notes on the pages to change the recipe depending on your oven or who's coming to dinner. Interestingly, the environment can alter the activity of genes, similar to making notes in a cookbook to slightly alter the recipe. In the last few years, we have learned that these small changes are not mutations to the DNA, yet they can be passed from one generation to the next. This exciting discovery is called epigenetic inheritance, a topic I will discuss in a later section.

Gene Expression is the Flow of Genetic Information

DNA does not directly make proteins, instead, it stores information for the long run, much like a library stores information in books. You may check out a cook book explaining how to make your favorite pizza, but the book does not make the pizza itself. Gene expression is the flow of genetic information from DNA to mRNA to proteins. The information in genes is first transcribed from a DNA template to messenger RNA (mRNA). Transcription is similar to copying notes from your text book or from a lecture. Once mRNA has been made, the information is then translated into a protein in a large molecular complex called a ribosome. Every protein on the planet was made in a ribosome using mRNA. It's important to point out that DNA is not turned into mRNA and mRNA is not turned into proteins.

Early in the 1960s, it a gene was defined as a sequence of DNA coding for a specific protein. In recent years, as our knowledge of genetics has grown, the exact definition of a gene has changed. New findings have shown that a single gene can make several proteins. Also, some sequences of DNA do not code for proteins at all, instead they code for sequences of RNA, an extremely versatile molecule used to carry genetic information, help make proteins, and regulate gene expression.

Cells Regulate Gene Expression

You are not what you were 6 months ago. Seriously, almost every molecule in your body is turned over about every 6 months. Yet, here you are in almost the exact same form as you were 6 months ago, including the same memories, body size, skin color, *etc*. Your body is like a river, you never step in the same river twice. Although the river flows in roughly the same place each year, the water is never the same just like molecules continuously flow through you. How does your body turn over almost every molecule within months, but you remain the same? The answer lies in how DNA maintains information and our cells regulate gene expression maintaining our bodies for decades.

In each one of us, there are more than 10 trillion cells forming about 200 different cell types from skin cells to nerve cells. The activity of genes in each cell must be precisely regulated because cells do not need to express all their genes at the same time. Not only would it be terribly inefficient, but your cells couldn't differentiate into different cells and organs making it impossible for us, or any type of plant or animal to exist. Each cell type expresses different proteins at different times, which explains why skin cells look so different from a muscle cell.

The ability of cells to regulate gene expression allows for rapid response to changes in their environment and not waste energy or resources. Recall that life must acquire energy and resources from its environment. However, environments can quickly change when new resources become available and others go away. Although individual organisms cannot adapt to their environment, they can quickly react to changes by fine tuning the expression of their genes to produce certain proteins when needed and turning off the production of others when they are not needed. Much like a volume knob on a radio, genes can be completely turned off, or ramped up to produce proteins at full volume.

To understand how cells can rapidly respond to their environment, imagine the billions of bacteria in your gut, they depend on the foods you eat for their survival, which can change with each meal. For example, if you drink milk, it floods your gut with lactose. To use this resource, bacteria make protein enzymes that break down the lactose. However, you may not drink milk all day, perhaps once or twice a week. To conserve energy, the bacteria will stop producing the enzymes needed to break down lactose when it is not present and make it again when you drink milk. They do this by turning transcription on or off, which prevents the lactose genes from being expressed.

Bacteria use operons to regulate gene expression. An operon consists of a DNA segment with the genes to make a protein along with the genes required to regulate the activity of that gene. Operons allow prokaryotic cells to rapidly respond to changing environmental conditions by ramping up gene expression or dialing it down, depending on the its environment

One major difference between prokaryote and eukaryote genomes is size, eukaryotic genomes can be hundreds to thousands of times larger than prokaryotic genomes. Prokaryotic genomes are optimized for quick reproduction with little non-functional DNA, whereas eukaryotes may have upwards of 80% of their genome that *appears* to not really be doing anything that we know of. Researchers are currently investigating why eukaryotic genomes are so large, whereas only a small portion, less than 5% of the human genome may actually code for proteins, and that is likely an over estimate.

At the start of the human genome project in the 1990s, scientists set out to sequence the entire human genome, a monumental task for the time and estimated to take over two decades and costs several billion dollars. By the end of the decade, the human genome was sequenced for a fraction of the originally estimated cost. When sequencing began on the human genome, friendly bets were made predicting the number of genes they would find in humans.

Original estimates were placed around 100,000 genes; after careful analysis of the human genome, that number has been revised downward to about 20,000 genes! Having so few genes has puzzled scientists and is a remarkable achievement of evolution considering that there could be upwards of a million different proteins produced by our cells. Although, more conservative estimates place the number of proteins closer to 100,000

Scientific discoveries almost always lead to more questions and then new discoveries. Once it became clear there were far fewer genes in our genome than predicted based on the number of proteins we potentially make, scientists began to ask how does our genome make so many proteins with fewer genes? By carefully analyzing the genes, another key difference was discovered between prokaryotes and eukaryotes. They both use the same genetic code, but eukaryotic genes have regions of intervening units called introns embedded within a gene. The part that actually codes for a protein is called an exon.

You can think of it like watching your favorite movie (the exons) interrupted by commercials (the introns). The purpose or origins of introns is not yet quite clear. However, it turns out that a gene can have many introns that are removed prior to translation, and there can be multiple exons within a gene. In a process called alternative mRNA splicing, cells rearrange the exons to make more than one protein from a single gene. In some ways our protein coding genes are similar to YouTube videos where people splice movies and television shows together into new stories by using old content. Just watch Star Wars vs Star Trek or Darth Vader vs Batman where splicing together different content creates unique stories.

New evidence suggests that some non-protein-coding segments of our genome make small non-coding RNAs to fine tune gene regulation. They can precisely control how many proteins are made and how long those proteins remain active inside cells before they become targeted for degradation. This ability to fine tune the regulation of genes is part of the reason why eukaryotes evolved into complex animals.

In the last decade, new technology has given us insights into how genes work with finer detail. For example, we know that cells use various mechanisms to regulate gene expression including operons, alternative mRNA splicing, and small non-protein coding RNAs. However, it turns out there's more to the story, gene expression can be regulated by process known as epigenetics, a word that means "above genetics". It works like this, small molecules called methyl groups can be added to the DNA where certain genes are located, effectively turning off that gene by preventing its expression. If you've ever seen a calico cat, the pattern of orange and black fur results from genes being permanently turned off. The same occurs in our cells, for example our skin cells do not produce the proteins that our eyes produce to detect light because those genes have been turned off.

We are just now learning that when our cells respond to the environment, those responses can be passed to our offspring through epigenetic inheritance. Some of the first evidence came from the Dutch Hunger Winter in World War II when the Germans cut off food supplies to the Dutch. As a result, women that were pregnant prior to the famine and gave birth during the famine had smaller children. This makes sense because babies gain most of their weight in the last three months of pregnancy. However, these children remained small throughout their lives. Even more remarkable, the grandchildren were also small. It turns out that starvation affected how genes were regulated, and it was passed down at least two generations. It is observations like these that are important so we can ask which genes are being affected by the Mom's environment, and how those patterns of epigenetic regulation are passed from parent to offspring.

Modern Advances in Genetics - 2000s

Modern Genetics began with Gregor Mendel in the 1860s and was greatly expanded in the early 1900s by Thomas H. Morgan. By the 1940s, the modern synthesis combined what was known about genetics with evolution by natural selection to create the paradigm of modern biology. One famous biologist, Theodosius Dobzhansky, once proclaimed that "nothing in biology makes sense except in the light of evolution". In the 1950s work by Rosalind Franklin, James Watson, and Francis Crick determined the structure of DNA and by the 1960s, the genetic code was cracked revealing the language of life. By the end of the second millennium, we had sequenced the entire 3 billion base pairs of the human genome. Today, we can sequence a genome in a few days for less than $2000, and the costs continue to decrease.

The history of life is written in our genomes. To understand that history, we have sequenced the genomes of bacteria, birds, mammals, fish, worms, and reptiles along with sequencing partial genomes of thousands of other organisms. By using sequences from many species, we know the relationships of living organisms, and develop models characterizing ancestral species. We have learned how genes are regulated in cells, quickly responding to environmental changes, or how other genes known as master regulatory genes, regulate the activity of other genes during growth and development. By understanding how these genes work we can learn how one cell differentiates during development, or how we grow limbs and organs, or the causes of cancer.

Modern genetics is also rapidly revolutionizing medicine as never before. By understanding how genes work, we are making huge strides in fighting cancer and why we age and die with the hope to treat cancer or slow down or even reverse the aging process. Advances in stem cell research may someday allow us to regrow damaged tissues, organs, or even new limbs that have been lost or damaged.

One major implication of the universality of the genetic code is that we can engineer organisms by inserting genes from one species into another, making genetically modified organisms (GMOs). While the use of GMOs became controversial among the public for various reasons, it has helped save the lives of thousands of people by allowing us to cheaply make drugs to cure diseases. One of the best examples of GMOs saving lives occurred when bacteria were engineered with human genes to make insulin to treat diabetes. People with diabetes often can't make insulin, so they require a constant supply of insulin or they will die.

In addition to engineering bacteria, scientists have engineered crops to improve yield and nutrition. Golden Rice was modified to produce beta carotene; a vitamin we require because our bodies can't make it. It's especially important for children because if they become severely deficient, it can lead to blindness or even death. Golden Rice could prevent thousands of cases of childhood blindness in the poorest regions of the world. Unfortunately, the countries that could most benefit by it have banned its import, wrongly believing it is bad or goes against nature. However, it's important to know that the beta carotene produced by Golden Rice is the same beta carotene produced by any other plant, including carrots. In this case, people's ignorance, ideology, and other wrongly held beliefs are cruel because they hurt people.

Recently, a new type of salmon was created by inserting the gene from another closely related species allowing the GMO salmon to grow year-round, thus reaching a larger size more quickly. This engineered fish is not bad for your health and will not hurt our health. In fact, most GMOs have undergone rigorous testing clearly demonstrating that they are safe for consumption.

New advances in CRISPR technology provide us a way to accurately and safely edit the genome of any organism. This gene editing technology is revolutionizing genetics and may be used to cure HIV, and inherited genetic diseases. It may also be used to stop the spread of diseases by introducing genes into the carriers, so they can no longer reproduce. For example, it has been proposed that we alter the genes of disease carrying mosquitoes making them sterile, which would greatly reduce their populations or even drive them to extinction, thus preventing the spread of malaria.

Scientific advancements in genetics is moving more quickly than society's ability to understand the implications of its findings. Therefore, it's important as a citizen to be informed about the implications of scientific advancement and make well-informed decisions based on facts and data. For example, there may be ecological consequences to driving mosquitoes to extinction. Luckily, disease carrying mosquitoes only count for a small fraction of all the mosquitoes. Losing one or two species of disease-carrying mosquitoes would most likely have little impact on the environment, but a disproportionately large impact on humans.

When it comes to humans, we should remove certain genetic defects, such as the ones that cause Tay-Sachs disease, Huntington's disease, or Achondroplasia. However, there are ethical concerns about editing the human genome for reasons other than repairing damaged genes. In 2018, a Chinese geneticists claimed to have modified the genes of twins to be resistant to HIV. This case brings to light the problems with modifying human genes. By making these kids HIV resistant, they may now be more susceptible to forms of encephalitis.

The debate about modifying humans will continue for decades. Some people will remain adamantly opposed to any changes, while others will want to alter people as much as possible to improve the species. In most cases, I think extreme positions harm society, we have a moral obligation to prevent human suffering, but the line between treating a lethal genetic disorder and enhancement can be a bit murky leading to an ethical gray area. Although, approaching the 2020's we still can't improve IQ, athletic or musical ability, or other such complicated traits. However, that day will come as we gain more information about how our genes work and interact with the environment.

Chapter 6

Reproduction Ensuring the Continuity of Life

You are the end result of an unbroken lineage going back more than 3.8 billion years to the dawn of life

Introduction

You are the end result of an unbroken lineage going back 3.8 billion years to the dawn of life. All the life we see today, including ourselves, can be traced back to an ancient common ancestor that was most likely a small population of bacteria living in the ancient oceans. For some, it can be hard to imagine that a small cell using a natural energy source to make organic molecules from carbon dioxide and hydrogen gas would eventually give rise to all the life we see today. Luckily for us, those primitive cells found a way to live independently of their place of origins and reproduce themselves.

Reproduction ensures that life continues on, slowly evolving into the myriad of species we see today. I have often wondered how many times life may have emerged in the first billion years of our planet, only to perish because it never acquired the ability to reproduce or leave its place of origins.

Reproduction can be relatively simple such as a prokaryote copying its DNA and dividing into two genetically identical offspring, or more complicated, such as a eukaryotic cell reproducing sexually by combining two genomes to form one offspring. The evolution of sexual reproduction and two sexes presented a paradox perplexing biologists for over a century. Quite simply, why would you combine two cells to form one offspring? And why cut in half the number of potential mates in a population with the evolution of separate sexes? Obviously, there are advantages to sexual reproduction, but teasing out its evolutionary origins has been difficult and filled with lively debates. I present a few of the theories in this chapter, some controversial, others more widely accepted.

With sexual reproduction comes sexual selection, a force of nature that has led to colorful ornamentation, elaborate dances, intricate songs, or enlarged adipose tissue surrounding mammary glands, all to attract choosy mates. Indeed, the need for reproduction has driven the evolution of some very strange looking animals and odd behavior.

Prokaryotes Reproduce by Binary Fission

The first cells reproduced by binary fission, a form of asexual reproduction. They copied their DNA, divided their cells roughly in half, making two genetically identical daughter cells. With each generation, mutations occurred when they copied their DNA. Natural selection acted upon this variation driving evolution in its random walk to an unknown future. Except for mutations, asexual reproduction produces genetically identical cells, which is a great way to reproduce as long as your environment remains relatively stable. You reproduce offspring that are fit to the same environment, over and over again.

Natural selection keeps prokaryotic genomes small, optimized for rapid reproduction when times are good. Copying extra DNA cost cells energy, time, and precious resources. Therefore, any unused genes, or extra segments of DNA, not serving a purpose are often lost after a few generations, maintaining a streamlined genome primed for rapid reproduction. Prokaryotes never combine two whole genomes into one organism, but they can exchange DNA between cells, even when they are not closely related.

Let's use the well-studied *E. coli* bacteria found inside your gut as an example of a typical bacterial genome. These tiny bacteria come in different strains where the number of genes range from about 4,377 to 5,416. The vast majority of these genes make proteins used to maintain the cell. But, you get the picture here, the average size of their genome is about 5,000 genes, give or take a few hundred. This translates to about 5 million base pairs of nucleotides. Natural selection constantly trims their genome to maintain rapid reproduction.

Within your gut, billions of *E. coli* make a living off the foods you eat. They aren't parasites by any means, instead, they are a vital component of our guts, aiding in digestion, producing beneficial molecules, and perhaps even preventing pathogens from causing us harm. Under favorable conditions, their population doubles every few hours, thanks to small genome size and rapid reproduction. Although, we refer to *E. coli* as a single species, a considerable amount of genetic variation exists between individuals. Some bacteria we call *E. Coli* may only share 20% of their genes. When you think about it, two bacteria of *E. coli* are more different than you are to a tree! But, what they all share similar metabolic pathways and genes required to let them thrive in your gut.

Prokaryotes Exchange Genes
Through Horizontal Gene Transfer

"It's complicated", one of my favorite responses to what seems like a simple question. The past 3.8 billion years of evolution leading to modern organisms, is "complicated". In the previous two chapters, I wrote that modern species evolved by natural selection where mutations create variation in our genes, which in turn are selected for or against by natural selection.

Over time, species change, reflecting changes in their genomes as they become better fit to their environment. But some genetic changes, including the evolution of new alleles or genes, were not created by a mutation in a parent and passed to their offspring. Sometimes, genes are acquired from the environment and incorporated into the organism's genome. Prokaryotes take in environmental DNA, random fragments of DNA floating around in their environment, which allows populations of prokaryotes to rapidly evolve in response to changing environments.

Using modern sequencing, we discovered that the pangenome of *E. coli* may include some 16,000 genes, a number that rivals the human genome. The pangenome of a population includes all the unique genes in a single population of one type of bacteria, not just a single bacteria. Vertical gene transfer occurs when genes get passed directly from parents to the offspring. But, prokaryotes also take up genes from the environment, a process known as horizontal (or lateral) gene transfer.

In horizontal gene transfer, bacteria acquire segments of DNA from the environment, adding these new genes into their own genome. These short sequences of DNA originate from closely or distantly related organisms, they don't even have to originate from other bacterial cells. This swapping of genes may not be reproduction in the strictest form, but it shows how living organisms operate in vastly different ways compared to what we are accustomed to in our own lives. In 2018, yet another discovery found that bacteria actually "fish" for fragments of DNA to incorporate into their genome.

Horizontal gene transfer also aids in the spread of antibiotic resistant diseases. Many types of organisms, including fungus and other bacteria, naturally produce antibiotics to protect themselves from bacteria. In turn, bacteria evolve ways to avoid being killed by antibiotics. Sometimes, harmful bacteria acquire genes responsible for antibiotic resistance through horizontal gene transfer and become highly selected for when antibiotics become common in the environment.

Prokaryotes can reproduce rapidly. It may be hard to believe, but there have been more generations of E. coli in your gut than there have been generations of humans since we last shared a common ancestor with chimpanzees about 5 million years ago! The combination of rapid reproduction, a large population, a large pangenome, and the ability of prokaryotes to incorporate environmental DNA, allow bacteria populations to adapt quickly to their surroundings when environmental conditions change. To complicate our concept of species and evolution even further, recent studies indicate that approximately two-thirds of the E. coli pangenome may have originated from other bacteria!

Horizontal gene transfer is possible because of the universality of the genetic code. It also allows us to manipulate the genes of any organism, including bacteria to create beneficial strains for research and medicine. It further complicates our simplistic view of descent with modification, because prokaryotes acquire beneficial genes from other bacteria without inheriting them from their parent cell in vertical gene transfer. Therefore, when we talk about our last universal common ancestor (LUCA), we almost always mean a population of prokaryotes rather than a single individual.

What exactly are we? A seemingly simple question with a complicated answer. It turns out that prokaryotes are not the only organisms with foreign genes. Eukaryotic genomes also possess bacterial and viral genes, and lots of them, starting from the wholesale transfer of mitochondrial genes that took place in the early evolution of eukaryotes. Today, human mitochondrial genomes have retained about 37 genes, just enough to regulate important protein activity for ATP production. That places about 500 genes in our nuclear DNA from the ancestor of mitochondria.

Other genetic studies have found nearly 100,000 viral fragments accounting for nearly five to eight percent of our genome. These ancient viruses inserted their genes into the genomes of our ancestors, only to be caught there and passed down through the generations. Luckily for us, mutations rendered most of them harmless.

Horizontal gene transfer in eukaryotes, including animals can strongly influence evolutionary trajectories. For example, the origin of mammals was likely caused by horizontal gene transfer from a virus! A unique feature of mammals occurs in development when the placenta embeds itself into the uterus of the mother. A special protein called syncytin makes this connection happen. Once researchers sequenced the gene for this protein, they immediately noticed that it matched the genes found in a virus. This means that a very mammalian trait, the placenta, has a viral origin, at least in part. It appears that horizontal gene transfer directly helped cause the evolution of modern mammals.

Eukaryotes Reproduce Asexually by Mitosis

During mitosis the nuclear material in eukaryotic cells gets divided and forms two genetically identical daughter cells. We often take for granted that mitosis includes an incredibly complicated set of coordinated actions carried out by the cell where everything must go exactly right. If not, the cell will die or even kill itself if it fails to line up and separate the chromosomes correctly.

Eukaryotes may be known for the evolution of sexual reproduction, but prior to that evolutionary innovation, they reproduced asexually. I imagine that early eukaryotes reproduced similar to binary fission; the DNA was replicated and then the cell divided into two genetically identical daughter cells. However, the extra ATP from mitochondria allowed for several changes to take place inside these cells in addition to becoming larger.

First, these cells used proteins to organize and regulate gene expression as the genome began to grow. Second, the length of the DNA grew larger allowing for more genes. And lastly, the DNA was cut into multiple linear segments forming distinct chromosomes unlike the single, circular chromosome found in prokaryotes.

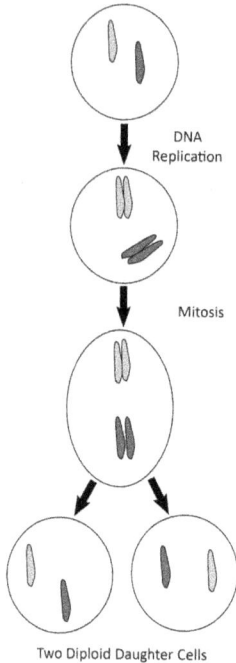

DNA Replication

Mitosis

Two Diploid Daughter Cells

Mitosis is asexual reproduction in eukaryotes. This cell has two copies of a single chromosome.

When eukaryotes reproduce asexually, the DNA is replicated in a similar manner to prokaryotes. However, the replicated DNA in eukaryotes condenses into visible chromosomes seen when dyed under a light microscope. Technically, prior to cell division, each copy of the chromosome is called a chromatid with both copies forming sister chromatids held together by glue-like proteins.

You can think of sister chromatids as two copies of a single chromosome comprised of DNA organized by specialized proteins. As we learned in the last chapter, the same segments of DNA always end up in the exact same place every time DNA condenses into chromosomes. It is this order that allowed early geneticists like Thomas Hunt Morgan to make genetic maps showing the locations of genes on chromosomes. What this means is that each of our 20,000 genes has a unique place in one of our 23 pairs of chromosomes.

Once the replicated DNA is tightly packed into chromosomes, the cell uses proteins to divide the cell in half. While there are many important proteins required for mitosis, I'm going to focus on microtubules, a type protein that forms part of the cytoskeleton of cells. Similar to our skeleton, a cell's cytoskeleton provides support and helps to maintain its shape. The microtubules resemble hollow tubes, like a pipe, and the cell uses them to organize the chromosomes during cell division. If the cell successfully moves the chromosomes to the center of the cell, then it sends a chemical signal to cleave the proteins holding the sister chromatids together. At this point, the replicated chromosomes move to opposite ends of the cell, pulled apart by motor proteins moving along the microtubules. When the chromosomes reach the opposite ends of the cell, the microtubules dissemble and the chromosomes begin to unwind from their condensed state. The last step involves pinching the cellular membrane in half forming two genetically identical daughter cells. We have spent over 100 years studying how the chromosome and, microtubules along with various other proteins coordinate their efforts during mitosis, and remains an active area of research.

Understanding mitosis has enormous implications for our growth, development, and health. All multicellular organisms are made of eukaryotic cells, and with the exception of our gametes, all the cells in our body reproduce by mitosis. Every day, you lose approximately a billion cells all replaced by mitosis. Recall from the last chapter that the activities of each one of these cells, including mitosis, must be under complete control. By learning how cells regulate mitosis, we could get new cells to grow when our tissues or organs become damaged. Imagine if we could grow new limbs for amputees or get our heart muscles to grow new tissue after a heart attack.

Unfortunately, there is a price we pay for being a multicellular organism, which is cancer. When our cells lose their identity and begin to divide uncontrollably, they become cancerous. Understanding mitosis helps in the fight against cancer. Have you ever seen someone on chemotherapy lose all their hair and their skin looks sickly? The reason why people lose their hair is because many cancer drugs work by halting mitosis. For example, the cancer drug Taxol prevents the separation of the sister chromatids during mitosis. If cells become stuck in mitosis, they will actually commit suicide through a process called apoptosis. An unfortunate side effect of chemotherapy is that it stops almost all your cells from dividing, that is why cancer patients have sickly looking skin, hair loss, and often feel nauseous.

The Evolution of Sexual Reproduction

Sexual reproduction is a uniquely eukaryotic process where two copies of a genome combine to form a new organism. Genetic recombination is the hallmark of sexual reproduction; however, its evolution and persistence among eukaryotes has puzzled scientists for over a century. First, based on common sense, you would think that asexually reproducing organisms should out-compete sexually reproducing organisms because asexual organisms give rise to their own offspring without the need to find a mate. Second, asexually reproducing organisms can produce offspring more quickly than sexually reproducing organisms. And third, another apparent disadvantage of sexual reproduction is that any parent would only be 50% related to its offspring rather than 100% related. Sexual reproduction also breaks up favorable combinations of alleles in an individual, potentially producing slightly less-fit offspring. However, within the past decade, new research has shed light on the age-old question of the origins of sexual reproduction.

The mitochondrial theory for the origins of sexual reproduction implicates endosymbiosis and the origins of our good friend the mitochondria as an evolutionary driver of sexual reproduction. Imagine a scenario where you are living inside your host, which you have become totally dependent upon as you've begun to lose most of your genes to your host. But, that's not a problem, the host provides you with all the raw materials you require. In return, you produce lots of ATP, that all important energy currency for the cell. Everything is going well, except that by using oxygen to produce lots of ATP, you also generate free radicals damaging your host's membranes, proteins, and DNA. Free radicals have unpaired electrons making them highly reactive, they want to quickly form bonds with any nearby molecule.

In response to free radicals, it is thought that eukaryotes evolved a nuclear envelope to house and protect the DNA. But alas, the cell's DNA can still be damaged. If enough damage occurs, the host cell dies, taking the mitochondria with it. If your host only has one copy of its genome and it becomes severely damaged, that is not good for you or your host. Now, imagine if you could somehow bring in a second copy of a genome? If it's not an identical copy from a close relative, then it would be less likely to have the same damaged genes. Also, with two copies of a genome, DNA repair mechanisms could fix the damaged DNA on one copy by using the other as template, guiding the repairs.

Recall that another paradox of sexual reproduction, it breaks up favorable combinations of alleles, or even produces unfavorable combinations. This problem was brought up when a beautiful movie star propositioned a famous comedian suggesting they should have children together. She told him that their children could have his brains and her good looks. But, the comedian replied, "They could also have my looks and your brains", a very possible outcome of sexual reproduction!

A second theory proposes that sexual reproduction can rid populations of deleterious alleles, which are versions of genes that lower the chances of successful reproduction. Some scientists argue that random mutations will eventually cause less-fit alleles to build up in a population. A single or even a couple of these slightly bad alleles may not be enough to prevent an organism from surviving and reproducing. In a population, these deleterious alleles may be spread across many individuals affecting the entire population. In sexually reproducing organisms, genetic recombination could put several of these deleterious alleles into one unlucky individual, basically rendering it unable to reproduce. In this case, the population would instantly rid itself of these undesirable alleles, thus improving the overall fitness of the population. You could imagine a Beavis and Butthead scenario where those kids would be unlikely to reproduce, so their less-fit alleles would not be passed on and be gone from the population.

Admittedly, this theory has received some harsh criticisms because it implies that sexual reproduction may be better for a population rather than providing direct advantages to the individual. However, some experimental data does support this theory. But, most experiments on evolution strongly support that natural selection operates on the individual, and it can be difficult to find evidence supporting natural selection acting on populations.

A third theory predicts that sexual reproduction increases genetic variation among offspring improving the chances of reproduction, especially in a variable environment. Sexual reproduction does not make new alleles, only mutations or horizontal gene transfer can serve as source of new alleles or genes in populations. But, sexual reproduction does create novel combinations of alleles that may work well together, the opposite of removing deleterious alleles from a population.

The evolution of sexual reproduction may have been driven by a combination of several factors where each improves the chances of reproductive success in the individual and improves the fitness of the overall population. The last two theories would be like playing Texas

Hold Em, some combinations of cards are winners while others are losers. For example, an off-suit 2 and 7 split, a losing hand would be like Beavis and Butthead, or an Ace, King, Queen, Jack, and Ten of Spades would be a winning hand like Einstein.

For whatever reasons sexual reproduction evolved, to combine entire genomes into an organism required a new way to reproduce. A recurrent theme in evolution is that it rarely re-invents the wheel, rather it modifies what's already there. In this case, sexual reproduction evolved by modifying mitosis.

Meiosis Increases Genetic Diversity

Sexual reproduction produces a new organism by fusing a male gamete (sperm) with a female gamete (egg) and then combining their DNA together. If gametes were created by mitosis, then the number of chromosomes would double each generation. Therefore, meiosis evolved to produce gametes with only one copy of the genome. Recall that the eukaryotic genome is chopped up into multiple linear segments called chromosomes with each cell carrying two copies of the genome. For example, humans have 46 total chromosomes, 23 pairs inherited from our parents.

Meiosis almost certainly evolved from mitosis with just a few small changes. Unlike mitosis, meiosis makes four genetically different gametes by using two rounds of cell division. The term ploidy refers to chromosome number. Eukaryotes are typically diploid (di=two), meaning they have two copies of their genome. In humans, we have 46 chromosomes in all the cells of our body. Whereas haploid means one copy, so our gametes possess 23 chromosomes. Our chromosomes are numbered from 1 to 22 based on size with the X and Y chromosomes being the sex chromosomes. Homologous chromosomes have the same genes in the same place. Humans have 22 homologous chromosomes, meaning you have two number 1 chromosomes containing the same genes, but potentially different versions of those genes (alleles).

Prior to cell division in any eukaryote, the DNA is replicated and condenses into chromosomes. In mitosis, two copies of each chromosome are held together forming sister chromatids, which are pulled apart during mitosis. But, meiosis does something a little different. Prior to the first cell division, the sister chromatids chromosomes join to form a single large structure known as tetrad made of four chromatids.

What happens in the tetrad is very important for inheritance because it generates new combinations of alleles not seen in either parent. Recall that Thomas H. Morgan showed that genes located on the same chromosome are inherited together. In humans, a great example of linked genes is red hair, green eyes, and fair skin. However, a process called crossing over occurs prior to the first cell division when sister chromatids of the homologous chromosomes fuse together into a tetrad.

It is here that different alleles from your parents can be swapped, forming new combinations on a chromosome. This occurs because homologous chromosomes carry the same genes in the same location, but within a population, there may be different versions of those genes (alleles). The tetrad allows alleles found on your Mom's chromosomes to jump to your Dad's chromosome and vice versa, creating new combinations on a chromosome. Going back to the common combination of red hair, green eyes, and pale skin, crossing over may lead to a person with red hair, dark eyes, and dark skin, although this is rare.

Meiosis also generates new combinations of alleles by a process called random assortment of the homologous chromosomes. During the first round of cell division in meiosis, the tetrads line up in the middle of the cell and are pulled apart, similar to mitosis. When the tetrads line up in the middle, their orientation is random, meaning the daughter cells receive a random assortment of maternal and paternal chromosomes. The number of possible combinations of maternal and paternal chromosomes grows exponentially 2^n, where n=number of chromosome pairs. If an organism has a total of four chromosomes (2 pairs), then the number of combinations is 2^2 (2 X 2) = 4, if there are three pairs then the number of combinations is 2^3 (2 X 2 X 2) = 8. In humans the number is much higher, 2^{23}, or just shy of 8.4 million possible combinations.

The first round of cell division in meiosis separates the homologous chromosomes, but the sister chromatids remain together. The second round of meiosis separates the sister chromatids, forming a total of 4 haploid daughter cells that mature into our gametes. In the act of sexual reproduction, two gametes randomly fuse forming a zygote, which grows into a new organism. In humans, there are approximately 8.4 million possible combinations of paternal and maternal chromosomes in each gamete, when they fuse that creates nearly 70 trillion possible combinations of chromosomes between the two parents. To put this another way, because of random fertilization, two parents would have to produce more than 70 trillion children before they produced two with the same set of chromosomes! That's a lot of potential diversity among offspring.

To sum it up, there are three ways that meiosis generates genetic diversity; crossing over between homologous chromosomes in the tetrad, random assortment of the homologous chromosomes prior to the first round of cell division, and random fertilization of the gametes. These three sources of variation create an incredible amount of diversity by shuffling genes around. It's much like playing poker with a 52-card deck. With each round, the same cards are shuffled and dealt, but the players are dealt new hands each new game. Different combinations of cards are more likely to win than others. It's the same with different combinations of alleles, some are more likely to be successful.

Why Have Two Sexes?

If sexual reproduction perplexed scientists, having two sexes was even more puzzling. Why would a system of reproduction evolve two distinct sexes where you immediately cut in half the number of potential mates? Why not just have everyone be the same sex, that way you would have an easier time finding a potential mate, reducing the probability of being out-competed by asexually reproducing organisms. Despite the potential shortcomings, sexual reproduction using two different sexes must impart some advantage because it is quite common among plants and animals.

To answer the origin of two sexes, we can once again turn to the mitochondria who seem to be responsible for many of the uniquely eukaryotic traits. If you live inside your host cell, and it becomes damaged, it could be beneficial to bring in a second copy of a genome. With two copies of the genome, you are likely to have at least one copy of a gene that works. If one copy of a gene becomes damaged, it becomes easier for DNA repair mechanisms to find and fix any damaged DNA.

But, how do you bring in a second copy of a genome? The easiest way is to fuse two cells together, like when a sperm fertilizes an egg. Meiosis evolved to produce haploid gametes, so that the number of chromosomes does not continually double with each generation. However, the two gametes are not equal in their size, or contribution to the next generation. In most eukaryotes, you can easily determine the female from the males by simply looking at the gametes (except for fungi, they do things a little differently, but I won't get into that here). In most cases, females produce fewer large eggs, whereas males produce many tiny sperm. Not only are the eggs larger, they contain all the organelles and mitochondria, whereas the sperm are usually tiny stripped-down cells carrying a copy of the father's genome. Basically, male gametes are nothing more than a source of genes.

Why do we inherit our organelles and mitochondria only from our moms? If you're a mitochondria living and reproducing asexually inside a cell, then all the other mitochondria are 100% related to you. There may be hundreds, or event thousands of genetically identical mitochondria in a single cell. If another cell were to fuse with your host bringing in new mitochondria, you would not be related to each other and would be competing against each other for resources. The evidence suggests that the evolution of two sexes was likely driven by mitochondria not wanting to compete with unrelated mitochondria from another cell. To point out the obvious, this system works because two sexes is practically ubiquitous across all eukaryotes.

Sexual Selection

Sexual selection has led to acrobatic dances, bizarre displays of brightly colored feathers, and other exaggerated features requiring energy to grow and maintain and practically scream to predators, hey come and eat me! Sexual selection occurs when one member of the other sex chooses a mate based on features that may not impart a direct survival advantage. Examples of sexual selection abound in the animal kingdom, ranging from the subtle, such as a small orange spot on a fish fin to the practically absurd of peacock feathers, all in the name of attracting mates.

Despite the energy costs of fancy ornamentation or dancing in the open, sexually selected traits have repeatedly evolved in many animals, often with spectacular examples. One theory to explain why sexually selected traits exists is based on the good genes hypothesis. It goes something like this; brightly colored feathers indicate that you have good genes and are otherwise healthy. It also means that you are well-fed, mostly disease and parasite free, and able to avoid predators even though you stand out. Maintaining colorful ornamentation or conducting fancy dances lets members of the opposite sex know that you are a good potential mate. In recent years, scientists have studied both female and male choice. Although female choice has been the most popularly studied, one species we are all familiar with, has undergone obvious male choice as we shall see.

Female choice is a type of sexual selection when females choose males based on certain traits that can seem arbitrary and random as fashion trends in our society. For example, in the 1980s neon colors were all the rage, but by the 1990s faded blue jeans and plaid shirts became popular. Just like our own society female choice in animals has led to some unusual behaviors and colorful ornamentation that serves no purpose in survival.

As a birder, I'm drawn to colorful feathers of birds, which provide a wonderful example of female choice. Much of the bright coloration in birds comes from females choosing males with the most attractive colors. In most birds with brightly colored males, the females remain drab compared to their male counterparts, mostly because it's in their best interest to remain somewhat camouflaged while incubating eggs on a nest or avoiding predators during the day.

Perhaps the best-known example of nearly run-away sexual selection occurred in the Indian Peafowl with their bright colors and spectacular tail feathers longer than the bird itself. All their feathers are beautifully colored in many shades of blue and green, they even have small bright blue ornamental feathers on their head. These brightly colored feathers don't necessarily help males with their day to day survival, but they do help attract the ladies! Think about it this way; if you're a male peafowl and can avoid predators, find food, and maintain those beautiful feathers, then you're a stud and the females will choose you because of your good genes. I should point out that the very large tail of peafowl may reduce predation by literally scaring away potential predators.

Sexual selection includes more than choosing brightly colored males, in some songbirds, females select males based on quality of their songs. In Northern Mockingbird, the males and females look identical. However, the males sing intricate songs known to include the various sounds in their environment. Male birds begin to learn their songs while still in the nest and continually add new sounds to and improve their songs throughout their lives. One particular mockingbird in north Florida learned how to mimic the author's alarm clock, which it liked to repeat every morning at the crack of dawn and sometimes in the middle of the night.

For some female birds, singing isn't enough, they like the males to sing and a display of beautiful feathers. Perhaps the greatest bird mimic is the Superb Lyrebird found in Australia. It can faithfully reproduce almost any sound in its environment including other bird calls, clicks from cameras, or even nearby chainsaws. While singing, it displays colorful feathers for admiring females. On lakes in western North America, female Western Grebes do not rely on fancy feathers or complex songs to find a mate. Their courtship relies on the male to dance in unison with the female while running on water.

Yet, for some female birds, they want it all: beautiful feathers, singing, and dancing. In prairie chickens, found in the grasslands of the Midwest, the males gather together in large groups called leks. When lekking, the males perform intricate dances, sing, and display their feathers, all to attract a mate. I imagine it is tough being a male prairie chicken having to maintain so many different sexually selected traits to attract a female, all while avoiding predators.

Female choice is not limited to birds. In darters, small freshwater fish found in clear streams, the males have colored fins with varying degrees of reds, blues, and greens that females must find appealing. In new world, male lizards in the genus Anolis use a colored throat patch called a dewlap they display to the females and other males. For some web-building spiders, the males are much smaller than the females and live in the corners of the web. When the time is right, they approach the female with great care, with each step perfectly choreographed. One miss-step and she will instantly pounce and eat him.

While examples of female choice are easy to find in nature, there are fewer examples of male choice, but there are some obvious ones close to home. It's well documented that animals spend a lot of time and energy attracting mates, but you don't have to go to exotic places to watch this phenomena, just go to your local night club and you will witness ornamentation, posturing, and sexually selected traits on display. Humans, like other animals evolved sexually selected traits and one of the best examples of male choice are female breast. The ancestors to modern male humans most likely chose enlarged breast because it was as sign of good genes and fertility. However, females don't need adipose tissue surrounding the mammary glands to make milk. In addition to enlarged breast, men also selected for fuller lips and a certain hip to waist ratio.

Indeed, human males have evolved to be visually attracted to certain traits of female bodies and the science fully supports this observation. When men look at attractive women, or are shown pictures of breasts, regions in their brain associated with rewards become strongly activated. Additional studies have also revealed that men score lower on IQ tests when pictures of attractive women are present because they it distracts them making it hard to concentrate.

Malaria
A Simple Organism with a Complex Life

Most of us are familiar with sexual reproduction, a male and female produce gametes, they get together and mate, so the gametes can fuse forming a new organism. However, not every organism uses such a simple life cycle. In fact, many parasitic organisms evolved complex life cycles with one or more stages reproducing asexually and another stage reproducing sexually. To complicate the life cycle even more, some parasites need more than one host to complete their life cycle, while using different reproductive strategies in each host. In most cases parasites reproduce asexually in their secondary host and reproduce sexually in their definitive host.

Malaria, one of the greatest health threats in the world, provides a great example of a parasite using a complex life cycle. Luckily for those of us living in North America, cold winters prevent the spread of malaria and in the south, we eradicated this disease by intensive insecticide spraying. Malaria is caused by a single-celled eukaryote in the genus *Plasmodium* and spread by mosquitoes in the genus *Anopheles*. When infected, the symptoms include fever, headaches, and vomiting, which can lead to coma and death. It takes about 10-15 days for the first symptoms to appear and recurrences can occur months later.

Despite the suffering caused by malaria, the life cycle of *Plasmodium* is quite interesting. It provides one example of a totally different way to reproduce and that humans are not the center of the animal world. I encourage you to read through this a couple of times, the terminology may be new, but focus on the overall picture. *Plasmodium falciparum* is one species that causes particularly severe cases of malaria. Similar to other eukaryotic organisms it reproduces sexually (meiosis) and asexually (mitosis). However, *Plasmodium* has several different forms in its life cycle depending on which host they are in

I'll start the life cycle of malaria with the infection of a person. It begins when a female mosquito bites, allowing the *Plasmodium* parasite to enter the blood stream. The parasite goes to the liver where it reproduces asexually by mitosis inside the liver. Over a short time, it produces thousands of clones that exit the liver and enter the blood stream. Once there, they begin to infect red blood cells where they undergo a second round of asexual reproduction further increasing their numbers and infecting more red blood cells. This is important because the higher frequence that red blood cells become infected with *Plasmodium*, the more likely they will get into another mosquito.

Humans are the secondary host for *Plasmodium*, we only harbor the parasite for a short time where it reproduces asexually to improve its chances of reaching the next stage of its life cycle. It may be hard to believe this, but we are basically disposable to the parasite. We are just there to ensure that the parasite gets into a female mosquito.

Once the parasites enter into another female mosquito, they

reproduce sexually by fusing to form a diploid zygote, or a fertilized egg. Once the zygote is formed, it will then divide by meiosis to form a haploid version of the parasite that will migrate to the saliva of the mosquito to be transmitted to the secondary host. And the cycle continues.

The female mosquito is the definitive host for malaria, which is defined as the host where sexual reproduction takes place. To the *Plasmodium* parasite, the survival of the mosquito is more important than the survival of humans, the secondary host. As you can see, humans are less important to *Plasmodium* than the mosquito, that's why humans infected with the parasite often die because it's in the interest of the mosquito to cause a higher rate of infection to ensure passage to the final host. Mosquitoes typically don't suffer from the *Plasmodium* parasites.

The Birds and the Bees
It is a Wild World

I end this chapter with my favorite examples showing the variety of sex determination and mating strategies in the animal kingdom. I named this section after an old saying about growing up, "it's time to learn about the birds and the bees." I've never understood this saying, especially as a biologist, especially because the way birds and bees reproduce is nothing like humans, except that some female birds prefer good looking males capable of singing and dancing. Let's begin with the bees because their reproduction is about as different from humans as you can imagine.

Male honey bees (*Apis mellifera*) do not have fathers, but they do have grandfathers. The same goes for ants, a close relative of bees, in fact you can think of ants as a wingless bee. As you may know, some bees and ants live in large colonies with a single queen laying eggs and many female workerss that do not reproduce. The males are drones with one function, mate with the queen. Here's how it works; female bees and ants are diploid, meaning they have a set of genes from the male and female, just like us. In contrast, males are haploid, with only one set of chromosomes from their moms, the queen. When the queen lays a fertilized egg, it grows into a diploid female worker; when the queen lays an unfertilized egg, it grows into a haploid male offspring. In this system, known as haplodiploidy, the male drones are 100% related to the queens and the female workers are 75% related to each other compared to human siblings that are on average 50% related to each other. Evidence suggest that the close relatedness of the colony allowed them to evolve complex social structures, which is rare in the animal kingdom.

"Stop parthenogenesis, have sex", pronounced a t-shirt one of my undergraduate friends used to wear. Parthenogenesis, the ability to reproduce asexually, has independently evolved in most animal groups except mammals. In the southwest, the entire population of New Mexico Whiptails (*Aspidoscelis neomexicana*), a ground-dwelling striped lizard, are females. In this species, the females have forsaken sex by getting rid of the males and reproduce asexually making genetically identical female offspring. The eggs remain unfertilized, yet grow into adults. But, there is a catch, these female lizards have to simulate copulation in order to induce egg formation. Lizards aren't the only animals reproducing by parthenogenesis. Aphids, small insects that feed on plants, also reproduce mostly by parthenogenesis. By the time the females hatch from an egg, they are already pregnant with the next generation!

The evolution of parthenogenesis goes to show that evolution by natural selection does not evolve species towards a goal. Presumably, parthenogenesis evolves because of the immediate benefits of not having to find suitable mates, allowing for rapid reproduction. However, because many parthenogenetic species are relatively young, it supports the theory that these species are likely doomed to an early extinction without sexual reproduction generating genetic variability.

In humans, gender is set very early in development, but this is not so for many species of fish, especially in the tropics. For example, fish in the family Serranidae, which include grouper, a popular game fish, are all born female. As they grow larger, some will mature and become males. Some species live in small groups where the largest female changes sex to become a male. Approximately 75% of fish that switch sex follow this pattern of being born female and later turning into a male.

The clownfish, popularized by Disney's movie *Finding Nemo*, also goes through a sex change during life. Small groups of male and female clownfish inhabit sea anemones and are immune to its stinging tentacles. Males are typically smaller than the female. When the female dies, the largest male grows larger and becomes a female. Imagine Disney having to explain to young children how Nemo's Dad became his Mom!

The Bluehead Wrasse (*Thalassoma bifasciatum*) complicates the picture even more. They can begin life as a male or a female, but it is the females that when they grow larger change sex. Not only that, they also go through several phases of life known as the initial phase and terminal phase. Larger females will switch to males and enter the larger terminal phase and hold territories for spawning. The smaller initial phase males possess larger testes than the terminal phase males, so they can compete with the larger males even though they do not guard a territory. To reproduce, the smaller males sneak in and quickly deposit their sperm when the larger males are spawning. Known affectionately known as sneaker males, they show you don't have to be the largest or strongest male to be successful at mating.

As we've seen in fish, some animals switch sex based on environmental changes. In humans, sex is determined by the two sex chromosomes, the X and much smaller Y chromosomes. XX means you are female and XY means you are a male. Other animals including some birds, fish, and reptiles have a Z and W chromosome where males are ZZ and females are ZW, the opposite situation of mammals. We learned about haplodiploidy in ants, bees, and wasp where males develop from unfertilized eggs and are haploid, whereas females develop from fertilized eggs and are diploid. In crocodiles and alligators, sex is not determined by sex chromosomes at all; rather the temperature at which the eggs incubate determines their sex. Eggs in warmer parts of the nest develop into males, whereas eggs in cooler regions of the nest develop into females.

Now for something even more complicated. A group of bacteria parasites in the genus *Wolbachia* infects many insects. Although, in some cases the relationship may be more mutualistic with both organisms benefiting from each other's presence. For example, some species of insects cannot even reproduce without the bacteria being present. Most types of *Wolbachia* are transmitted to the next host through the eggs of the female. To increase their transmission rate to their next host, the bacteria has been known to kill males while still larvae or turn males into females, and induce parthenogenesis in females.

There are many other stories here; in fact, careers have been built on studying how life reproduces from understanding the genetic bases of sex to the evolution of different behavior and mating strategies across the species. But, despite the enormous variety of behaviors and sex determination, one thing remains, sexual reproduction with two sexes evolved in our distant eukaryotic ancestor and has not only been maintained, but has diversified as well.

Horseshoe crabs have returned to sandy beaches for 450 million years to lay their eggs at the edge of the water during spring high tides

Chapter 7

The Unseen World
of
Single-celled Organisms

In a single spoonful garden soil, there are more
bacteria than there have been people that have ever
lived on this planet.

With 3.8 billion years of evolutionary history,
don't think for a second prokaryotes are primitive

Introduction

The first fully independent living organisms on Earth were structurally simple prokaryotic cells, for life started out simple by necessity. Indeed, the name prokaryote means "before the nucleus", an accurate description as prokaryotes existed for some 1.5-2 billion years prior to the arrival of the larger and structurally more complex eukaryotes. Even today, prokaryotes still dominate the Earth in sheer numbers and perhaps biomass too.

When it comes to diversity, prokaryotes get passed over due to their seemingly simple structure, making it difficult to tell different species apart by simply examining them under a microscope. However, we are in a golden age of genetic discovery brought about by rapid DNA sequencing techniques. For the first time, we can search for new bacteria by sequencing tiny fragments of DNA from bacteria living on a grain of sand, or on small sample of human tissue. The findings have been nothing short of remarkable, we now know that prokaryotes are everywhere and much more diverse than we ever thought. While appearing structurally simple, prokaryotic diversity lies in their metabolism where some extract energy from rocks, thermal vents, and sunlight, while others can live in boiling water or in battery acid.

It may seem odd, but eukaryotes can thank their origins to the merger of two prokaryotes about 2 billion years ago. What ensued was the largest restructuring of cell complexity in the nearly 2 billion years life had existed as eukaryotic cells evolved internal membranes to compartmentalize various functions of the cell. This single event paved the way for multicellular life including plants, fungi, and animals. But, eukaryotes still relied on prokaryotes and continue to do so today. Plants, and animals form relationships with many types of bacteria. Despite, all the anti-bacterial cleaning products, we can't and don't want to live in a sterile environment free of bacteria.

Prokaryotes Paved the Way for Life

Life started out simply, because it had to. For nearly two billion years, life remained structurally simple, but the impact of single-celled prokaryotes to the Earth cannot be understated. Returning to a major theme, life needs energy gets energy from the environment. Over time, microbial evolution provided new ways to get energy from the environment by exploring new resources

Based on the fact that basically every cell uses proton gradients to make ATP, it is thought that the first cells harnessed the potential energy from proton gradients in alkaline vents. Once these early cells left their places of origins, they probably acquired energy from organic molecules in their environment. However, the most abundant and practically endless source of energy in the world is sunlight. It didn't take long for life to tap into this energy source and it's not surprising that photosynthesis evolved early in the history of life.

Photosynthesis is widely known for its crucial role in transforming the energy from sunlight into chemical energy stored in carbohydrates, thus making the energy in sunlight available to life. But, it also makes carbon available to life by fixing carbon dioxide into organic molecules. However, the impact of photosynthesis does not stop here. The byproduct of photosynthesis is free-oxygen (O_2), over billions of years the free-oxygen released from photosynthesis transformed the atmosphere into an oxygen rich mixture of gases that energized life.

Once atmospheric oxygen levels reached critical levels, the oxygen became toxic for many bacteria. But some prokaryotes evolved ways to deal with the toxic effects of oxygen by harnessing its electron grabbing tendencies to make more ATP. This one small act not only neutralized the toxic effects of oxygen by forming water, it also made much more ATP for the same amount of fuel. It was like going from five miles per gallon to over thirty miles per gallon without sacrificing any performance.

Because cells could make more ATP, oxygen energized these early cells, they were more efficient at extracting energy form their environment and therefore had more energy to work with. It was the evolution of aerobic respiration in bacteria that paved the way for eukaryotic cells and eventually multicellular life. In fact, the merger of two prokaryotic cells, an aerobically respiring bacterium and an archaeon that led to origins of eukaryotic cells.

In addition to energizing life, oxygen is also important high up in the atmosphere (10km to 50km) where it reacts with incoming solar radiation to form a protective ozone layer. Ozone is comprised of three oxygen atoms (O_3) and filters out harmful UV rays from the sun. If it weren't for the ozone layer, incoming UV rays from the sum would sterilize the surface keeping it barren of almost all life. The Earth's oceans would also risk disappearing because ultraviolet light breaks water into oxygen and hydrogen gas. Our planet isn't quite large enough to hold on to hydrogen gas, so it escapes into space. Without an ozone layer, the Earth's oceans could slowly disappear, making the planet a dry place, and not very habitable for life. Just look at the dry surface of Mars. It once had water on its surface, but it is now gone. Based on what we see on the Earth, some think that life never had a chance to stabilize the Martian atmosphere. We can thank photosynthetic microbes for stabilizing our atmosphere, protecting our oceans, and making it possible for multicellular life.

Prokaryotes are Difficult to Classify

Human nature has a need to classify objects into discrete units. Think about it this way, the borders of our country, states, congressional districts, and counties are clearly marked on maps. You live in New Mexico or Colorado based on which side of a line you reside. North of the state line, you live in Colorado, south of the state line, you live in New Mexico. There isn't a middle ground. But, to migrating birds, or elk living in the San Juan Mountains, the habitat is continuous, and our political boundaries are meaningless. When it comes to animals, we classify them into a nested hierarchy from the most inclusive of kingdom (similar to a country) to the least inclusive of species (a county). In fact, many biologists will argue that a species is unique and different from all other species.

However, nature often defies our attempt to neatly classify the variety of life into discrete units. Nowhere is the problem more prevalent than in the prokaryotes because they really don't fit into a nested hierarchy of kingdom, phylum, class, order, family, genus, and species.

One way to classify organisms is based on grouping them by evolutionary relationships of taxa, which we can illustrate as a phylogenetic tree. We construct these relationships based on morphological and genetic similarities. The more two organisms have in common, the more recent their common ancestor and the more closely related they are to each other. Once again, the prokaryotes make this task incredibly difficult due to their large pangenomes and ability to swap parts of their genome in horizontal gene transfers.

Prokaryotes reproduce asexually producing genetically identical daughter cells, with the exception of random mutations that introduces a little variability. But, as you learned in the last chapter, horizontal gene transfer runs rampant among unrelated prokaryotes and apparently regardless of how closely related they are to each other. While a typical E. coli may possess 5,000 genes, the entire pangenome may include 16,000 or more genes making it difficult to neatly classify them as a unique species. In animals it's a bit easier to classify species, but could you imagine routinely swapping genes with organisms as different as a tree or a mushroom? Yet, that is exactly what prokaryotes do all the time. There is more genetic variation in E. coli in your gut than there is between you and a lizard, and we haven't shared a common ancestor with a lizard in over 315 million years.

Carl Woese in the 1970s was among the first to realize the immense diversity of prokaryotes. At the time, most scientists divided life into five kingdoms (bacteria, protists, fungi, plants, and animals) with four of the kingdoms comprised of organisms with eukaryotic cells. Woese used advanced molecular techniques for the 1970s to study the similarities of ribosomes between the five kingdoms. Recall that ribosomes are used to make proteins in all cells. Their general structure is highly conserved because any change to its shape would almost certainly be harmful. But, over time, some small changes do occur and are passed down through the ages. Because ribosomes and the genes that code the information to make them remain highly conserved, they can be used to reconstruct evolutionary relationships of very different organisms that diverged millions to even billions of years ago.

The results of Woese's study showed that a relatively unknown group of prokaryotes living in extreme environments were vastly different from most other prokaryotes. He also discovered that there were far fewer molecular differences between the four kingdoms of eukaryotes, the protists, plants, fungi, and animals than there were among the prokaryotes. The results were ground-breaking, leading him to propose a new classification system consisting of three domains made of bacteria, archaea, and eukarya. Four kingdoms of eukaryotes were combined into one domain, while the deceptively simple bacteria were divided into two domains, bacteria and archaea. The new classification was based primarily on differences in the ribosomes. Despite the vast outward differences between plants and animals, at the molecular level, all eukaryotes are very similar.

Soon after Carl Woese's study, there was renewed interest in studying prokaryotes, and it became easier with new technologies making it easier and cheaper to sequence DNA. Additional discoveries showing the differences between bacteria and archaea continue to support Woese's study. To start, major differences exist between the structure of their cellular membranes. While they are both made of phospholipids, they use different building blocks requiring unique metabolic pathways to create their phospholipids. Using different building blocks to construct cellular membranes may not seem like much, but scientists studying the origins of life have speculated that the differences are a strong indication that life acquired their cellular membranes twice and independently; once for the bacteria and once for the archaea. This means that bacteria and archaea began to diverge at the origin of life itself.

Although we lump bacteria and archeans together as prokaryotes, it remains unclear just how many different types of prokaryotes exist. To further appreciate prokaryote diversity, check out these facts. Genetic studies analyzing the bacterial diversity in a handful of soil estimates upwards of 8 million types of bacteria. Studies of leaves from one species of tree revealed several hundred species of bacteria, which were different compared to the bacteria on the leaves of another tree. We already know that each one of harbors thousand species of bacteria living in our gut, and there may be thousands of different species of bacteria capable of living in our gut, and we are just one type of animal. To put this diversity into perspective there are only 5,400 species of mammals, about 10,000 species of birds, 300,000 species of plants, and about 2 million species of animals. Yet, a single handful of garden soil may house the same amount of prokaryotic diversity as all the plants and animals in the entire world!

Prokaryotes are Metabolically Diverse

Don't call prokaryotes primitive just because they lack a cell nucleus and remain otherwise structurally simple. What prokaryotes lack in structural complexity compared to eukaryotes, they make up for it in metabolism. Prokaryotes were the first organisms to evolve photosynthesis nearly 3.5 billion years, forever transforming the surface of the Earth by releasing free-oxygen to the atmosphere. In turn, this paved the way for aerobic respiration, the ability to use oxygen to vastly improve the efficiency of ATP production. Additionally, the origin of eukaryotes occurred when two very different prokaryotes merged together with one cell living inside the other in a mutually beneficial relationship. This unlikely merger ultimately paved the way for active multicellular animals.

Life is an island of low entropy, using a myriad of chemical reactions to create order by taking in energy and materials from the environment. The origin of life traces its roots back to chemical evolution beginning with the simple reaction of adding hydrogen to carbon dioxide, or fixing carbon into an organic molecule. Broadly speaking, organisms fall into one of two categories, heterotrophs (hetero=other, troph=feeding) or autotrophs (auto=self, troph=feeding). Heterotrophs acquire nutrients and energy from organic molecules, typically made by autotrophs. Autotrophs are self-feeders extracting energy from the environment to fix carbon into organic molecules, and then making all the molecules they need. Organic molecules are based on the element carbon because of its ability to form four stable chemical bonds allowing it form millions of different organic molecules.

There are only a few metabolic processes that add hydrogen to carbon dioxide and autotrophic prokaryotes evolved them all. Recall that metabolism is the sum of chemical reactions taking place in cells that bring them to life. Some reactions break down molecules into smaller ones, often accompanied with a release of energy that cells use to do work. In turn, the smaller molecules get re-purposed as the building blocks for larger more complex molecules. Next, I explore the microbial diversity in our world starting with chemoautotrophs, a type of organism that does not use sunlight to fix carbon dioxide.

Chemoautotrophs - Methanogens

Have you heard of an Archaean known as a methanogen? Don't feel bad, most people have not. Yet, they the most important organism to our evolution you never heard of. Based on current research, they may have been the cell that an aerobically respiring bacteria took residence in over 2 billion years ago, jump starting the evolution of eukaryotes! Methanogens react hydrogen gas (H_2) with carbon dioxide (CO_2) to make methane (CH_4), an organic molecule found in natural gas. This is an example of carbon fixation when inorganic carbon is converted to organic carbon by the addition of hydrogen atoms.

They say there is no such thing as a free lunch, but that's not entirely true all the time, at least for methanogens. Hydrogen gas has a lot of potential energy, the bonds are easy to break; it practically wants to react with other molecules. With a little help from a metal catalyst, hydrogen gas reacts with carbon dioxide making methane and it releases a little bit of energy. It's like getting paid to eat lunch! This means that reacting hydrogen with carbon dioxide is favorable, unlike most reactions involving carbon dioxide which require energy.

If you've heard of swamp gas, it's methane made by methanogens. In recent years, some methanogens have been found in some people's making methane, similar to swamp gas. The implication is they have flammable flatulence.

Chemoautotroph - Iron Bacteria

Did you know that some bacteria eat rocks for a living? Well, not exactly the same way animals eat, but they do get energy from rocks. To understand how they pull this off, recall that electrons can serve as source of energy, hence the connection between electricity and electrons. Even the electricity in your home is the result of electrons moving through the wires from the positive end to the negative end.

Some bacteria obtain energy from rocks by snagging their electrons. Specifically, a form of reduced iron, known as Iron 2 (Fe^{++}), donates energy carrying electrons used to fix carbon dioxide into organic molecules. In chemistry, reduced means you have electrons, whereas oxidized means you've lost electrons. If you've watched iron rust, it's slowly being oxidized, losing its electrons to oxygen. Prior to the oxygen revolution brought about by the photosynthetic action of cyanobacteria, much of the iron was reduced and able to donate electrons to iron bacteria. Iron bacteria rely on the reduced form of iron found inside of rocks, or rocks that haven't been exposed to the air for too long. In addition to iron, bacteria can also grab electrons from other minerals as well.

If you've ever been in the mountains, you may have seen bright orange streams coming from abandoned mines. Iron bacteria are responsible for the orange-colored acidic water, a major problem for stream quality. As mines age, they slowly fill with water. Bacteria and archaea begin to grow and oxidize (remove electrons) the minerals, slowly lowering the pH and oxygen levels of the water. These bacteria actually favor the acidic conditions and are a type of extremophile known as an acidophile. The bright orange color is the result of the oxidized minerals, mainly iron minerals.

Photoautotrophs - Cyanobacteria

Contrary to popular belief, plants did not evolve photosynthesis, it was acquired from bacteria. Photosynthesis uses the energy in sunlight to fix carbon dioxide into carbohydrates. Recall that every cell on this planet uses a proton gradient to generate the majority of its ATP. Through evolution, cyanobacteria were able to re-purpose the proteins involved in making proton gradients into photosystems that transform the energy in sunlight to chemical energy. The origin of plants began about one billion years ago when a eukaryotic cell engulfed a cyanobacteria. Rather than digesting it, the cyanobacteria became an endosymbiont along with the mitochondria forming the first eukaryotic algae. One lineage of algae went on to evolve into plants, which dominate terrestrial ecosystems.

Nitrogen fixers

The Earth's atmosphere is about 78% nitrogen (N_2), which is also one of the four most abundant elements found in living organisms. Despite being abundant in the atmosphere, nitrogen gas is inert and not very reactive making it practically unavailable to most living organisms. For life to use nitrogen, it must be "fixed" into molecules available to organisms, much like carbon must be fixed into organic molecules using energy.

Nitrogen fixation also requires an oxygen free environment and energy to fix nitrogen into ammonia (NH_3) or nitrates (NO_3). No eukaryote (plants or animals) can fix nitrogen, but bacteria can. However, plants in the pea family possess root nodules that house a symbiotic bacteria called *Rhizobia* that fix nitrogen into usable forms. The roots create an oxygen free environment and energy while the *Rhizobia* bacteria provide usable nitrogen to the plant.

Reverse Krebs cycle

For most biology majors, remembering the steps of the Krebs cycle is a rite of passage; a series of eight chemical reactions where bonds are broken and new ones are formed, thus creating new molecules at each step (although, the details soon forgotten after an exam). Recall that the Krebs cycle's purpose is to break down organic molecules by stripping them of their electrons, forming carbon dioxide as a waste product. The high energy electrons are used to make ATP for the cell.

Having taken any intro-biology course, you would think that the Krebs cycle exist to break down organic molecules as part of the cells way to generate ATP. However, if you look at any metabolic chart, you would see a circle representing the Krebs cycle in the middle of the chart. The intermediates made from the Krebs cycle can be siphoned off and used as the building blocks of lipids, carbohydrates, proteins, nucleic acids, and various other molecules required by the cell. The Krebs cycle is ubiquitous among all living organisms betraying its ancient ancestry to the first cells. Its central importance to cellular metabolism indicates it may have evolved through chemical evolution before the first cells had membranes.

But, what you may not know is that if you add energy to the Krebs cycle and hydrogen gas (H_2), it can add hydrogen to carbon dioxide! The Krebs cycle is a metabolic pathway that can fix carbon by adding hydrogen to carbon dioxide, meaning it can run backwards. Rather than being catabolic, only capable of breaking down organic molecules, with an input of energy, it can run backwards allowing the organism to become autotrophic. However, it's not well known how often this happens.

Prokaryotes Live in Extreme Environments

Prokaryotes are everywhere, including some of the most bizarre and extreme environments you could imagine. They are found thousands of feet below the surface, living in aquifers and extracting energy from rocks. They have been found living in water hot enough to boil, or in freshwater lakes thousands of feet below glaciers, or living in rocks perpetually frozen in the Antarctic. As you just read, iron bacteria live in very acidic places in mines as they slowly oxidize rocks for their electrons. Below, I describe a few examples of prokaryotes living in extreme environments.

Extreme cold

In the 1990s, a team of researches braving the cold and harsh environment of the dry valleys in Antarctica discovered bacteria eking out a living inside the frozen rocks. We usually think of the Antarctic continent as being totally covered in ice; however, there are a few places where precipitation is so low that the snow sublimates. Sublimation is when ice turns directly into vapor, and in the Antarctic it leads to the formation of dry valleys. The lack of precipitation combined with extreme cold makes this one of the harshest environments on Earth. Yet bacteria known as endoliths, often referred to as "rock eaters", survive here by stripping electrons from rocks as an energy source. A trade-off for such a harsh living is that these bacteria grow incredibly slow.

Scientific discoveries did not stop with finding bacteria in Antarctic rocks of the dry valleys. In 2014, scientists discovered other "rock eating" bacteria in a freshwater lake buried several thousand feet beneath Antarctic ice. Archaea and bacteria that live in extremely cold environments are known as psychrophiles. Astrobiologists study bacteria living in these extreme environments to gain insight into how life may exist on Mars, or other planets and moons, where it is also cold and dry.

Extreme salinity: The Dead Sea and the Great Salt Lake in Utah share a few things in common, but most notably, they lack fish, a direct result of extremely high salinity. Although, the name "Dead Sea" is not quite accurate, on the contrary, it is full of bacteria and archaea surviving in these salty environments. Known as halophiles, these microbes spend extra energy to keep water int them. One way to they do this is by making protectants that help keep water inside the cell. Without these protectants, water would rapidly leave the cell through osmosis causing them to shrivel and die. Another way of keeping water inside halophiles involves pumping of potassium ions into the cell. Ironically, if you were to place a halophile in freshwater, it would immediately swell and burst from the water rushing into the cell.

Extreme heat

Yellowstone is known for its geysers, including Old Faithful, along with many hot springs with water temperatures approaching the boiling point. Prior to the 1960s, it was generally assumed that life couldn't exists much beyond 135°F (55°C), then bacteria were found thriving in temperatures up to 160° even 175°F. One of these small unassuming thermophiles was named *Thermus aquaticus*, which literally means "hot water". An important adaptation for these bacteria are proteins capable of working at hot temperatures without falling apart. All proteins perform some task, and to do so, they have to be the right shape, changes in pH, temperature, or salinity will change the shape of proteins making them useless. If you've ever cooked an egg, you have seen how heat alters proteins when the egg white changes from clear liquid to a white solid. Those types of changes to proteins in a cell would render them useless leading to cell death.

The discovery of *Thermus aquaticus* turned out to be one of the most important discoveries for the field of genetics. Here's why. Every cell replicates its DNA using a large protein complex called DNA polymerase. In *Thermus aquaticus*, its DNA polymerase works normally under very hot conditions. This may not seem to be overly important; however, finding it led to a major advance in our ability to sequence DNA.

In the early 1980s, a PCR was invented to amplify small amounts of DNA making it easier to sequence. However, as part of the process, the DNA is heated and cooled repeatedly. Unfortunately, the heating process denatures most DNA polymerase used to amplify the DNA, rendering it useless. As a result, it was difficult to amplify small segments of DNA for sequencing. However, using the DNA polymerase from *Thermus aquaticus* solved the problem of using heat as part of the DNA amplification process and allowed scientist to easily amplify sequences of DNA.

The DNA polymerase from *Thermus aquaticus* is called *Taq* polymerase. Its patent was sold for $330 million dollars and is speculated to have generated over 2 billion dollars in royalties. Its discovery also led to the Nobel Prize in 1993. *Thermus aquaticus*, a small thermophilic bacteria living in a hot spring in Yellowstone National Park led to a revolution in genetic research. The discovery of *Taq* polymerase clearly shows the importance of finding, studying, and preserving biodiversity in the world because you never know what you may find!

Disease Causing Bacteria

The vast majority of bacteria are harmless, some are beneficial, and some may even be required for our existence. And, of course, some bacteria cause diseases. Unfortunately, pathogenic bacteria are responsible for several terrible diseases including tuberculosis, salmonella, tetanus, syphilis, staph infections, along with various types of food and waterborne illnesses. There are many bacterial diseases, but I'm only going to focus on two, tuberculosis and pertussis (whooping cough).

Tuberculosis is a highly contagious, airborne disease that has killed millions of people. In 2013, there were 9 million new cases of tuberculosis causing approximately 1.3-1.5 million deaths. Many new cases are in poor regions of the world lacking adequate medical care and access to vaccinations.

The bacteria *Mycobacterium tuberculosis causes* tuberculosis (TB), which infects the lungs causing major damage. Widespread vaccinations have dramatically reduced TB outbreaks in developed countries including the US. However, where vaccinations are less common, or people refuse to get vaccinated, TB remains prevalent. Most TB infections can be successfully treated with proper antibiotics. However, drug resistant strains continue to evolve, and spread around the world infecting people who have not been vaccinated.

Whooping cough, also known as pertussis, is a highly contagious airborne disease caused by the bacteria *Bordetella pertussis*. It was once widespread, killing between 10,000 and 20,000 people per year in the US, with higher mortality rates in infants and young children. Pertussis produces a toxin that stops the cilia in the cells that line your throat. As a result, people with whooping cough will go into major uncontrollable coughing fits. The first vaccine was developed in the 1930s and mortality declined to less than 100 by the 1980s.

With the accidental discovery of penicillin in 1926, survival rates against bacterial diseases and infections greatly improved. Combined with the widespread implementation vaccinations, some bacterial infections including diphtheria, tetanus, pertussis (Whooping Cough) and tuberculosis have continuously declined to the point where they have almost been eradicated. Unfortunately, these diseases are on the rise simply because people don't get vaccinated.

Vaccinations work because we have an acquired immune system, meaning it remembers every pathogen that has invaded your body, including ones that made you sick or weakened versions from vaccinations. By getting vaccinated, our immune system remembers the pathogen allowing it to quickly react to the bacteria or virus, eradicating it before it makes us sick.

Sadly, uninformed parents are not getting their children vaccinated against these serious diseases. They wrongly believe that vaccinations lead to autism or other forms of mental impairment. However scientific studies have resoundingly disproved the claim that vaccinations cause autism or any type of mental impairment. In fact, the *Journal of the American Medical Association* (JAMA), one of the most respected journals in the world called the connection between vaccinations and brain injury a "myth" and "nonsense". Just to be perfectly clear, VACCINATIONS DO NOT CAUSE AUTISM OR MENTAL IMPAIRMENT.

In 2015, an outbreak of whooping cough in California led state lawmakers to remove personal exemptions for the vaccinations for all children entering the public school system. Getting a vaccination only reduces your risk of contracting the disease and works best when everyone else around you is also vaccinated. If a large portion of the population refuses their vaccinations, they are at much higher risk for diseases and they put others at risk as well. Some studies have even shown that the effectiveness of vaccinations does not last a lifetime, but wanes over time, sometimes lasting less than a decade.

Infectious diseases will always remain a problem, they continue to persist because not everyone gets vaccinated, and because the pathogens evolve causing a rise in antibiotic resistant diseases. As I discussed in Chapter 6, horizontal gene transfer allows bacteria to swap genes with each other including genes for antibiotic resistance. Because of the widespread use of antibiotics, especially in the meat industry, antibiotic resistant bacteria are on the rise and spreading.

Beneficial Bacteria – Our Microbiome

Bacteria get a bad rap because they conjure up images and feelings of uncleanliness and diseases. The fear of bacteria is rampant, just walk down the soap aisle at the local store and you will find labels claiming how well their product kills bacteria. One brand of dish soap claimed to kill 99.9% of all bacteria. At first, this may sound really good. But, if there were 10 million bacteria on your spoon and 0.1% lived, you would leave 10,000 bacteria, and they would be more resistant to your soap! You just caused the bacteria in your kitchen to evolve to be resistant to your soap.

Fortunately, the vast majority of bacteria are harmless, you don't even know they are there. For example, a single spoonful of dirt contains more bacteria than there are people living on the Earth. In reality, bacteria are incredibly important for the functioning of our ecosystems and ourselves. Complex life and ecosystems could not exist without bacteria and archaea.

It turns out that animals depend on symbiotic bacteria for their existence. Your body harbors a lot of bacteria; they potentially outnumber your cells 10 to 1. For every one cell that is you, there are perhaps 10 bacterial cells living in and on you. It has been estimated you have about 10 trillion cells that are you, and about 100 trillion prokaryotes. For years, we have known that *E. coli* was the most common bacteria living in our gut, along with a few other types of microbes.

Recent studies have revealed that each one of us houses a diverse microbiome potentially more beneficial than we imagined a decade ago. The average person has over a thousand different types of bacteria residing in their gut alone, and it is their diverse metabolic activities where we get our benefits. We have also discovered that no two people have exactly the same gut flora. Studies are showing that environment, including diet and where someone lives, along with their genetics play an important role in determining the types of microbes present.

Currently, the full extent of the diversity of gut microbes in humans is not well known. In fact, their effect on our bodies is just starting to come into light. We have known that bacteria help us break down our foods and they also make various vitamins and enzymes we need if they are lacking in our diet. The sheer volume of some bacteria, including *E. coli*, may form a protective barrier in our gut preventing us from getting diseases. The influence of our gut microbiome doesn't stop here. In recent years, growing evidence suggests that our gut microbes influence our behavior.

In studies using mice, some microbes affect serotonin levels in the gut. Serotonin is a chemical used by the brain to relay signals to the body, this one molecule can affect mood, appetite, sleep, memory, and sexual desire. What this means, is that there could be a link between our moods, cravings, and health based on gut our microbes, which in turn are strongly influenced by our diet. Understanding how microbes influence us is only in its infancy and will continue to be a source of research for years to come.

In recent years, the probiotic industry has grown promoting various foods containing beneficial bacteria, or probiotics including those found in yogurt (although, they must have live active cultures). A serving of yogurt with live active cultures every day can have a positive influence on your gut microbe by promoting beneficial bacteria including several species of *Lactobacillus*. Having good bacteria can help a wide range of conditions including chronic constipation, diarrhea, and inflammatory bowel disease.

Observational studies on yogurt indicated that people who regularly eat yogurt have lower rates of obesity. In an effort to understand how yogurt and probiotics effect obesity, scientists conducted a controlled experiment on mice where some received yogurt with live cultures and others did not. At first, the scientists noticed that the hair of the mice on a daily diet of yogurt intake was shiny, coinciding with increased follicle activity, a clear sign of improved overall health. However, the researches made another unsuspected finding, they noticed that the males were "projecting their testes outward" giving the mice swagger. Indeed, the males that were fed yogurt had testes sizes 5% heavier than the males not eating yogurt. Additionally, they sired more offspring. Yogurt also paid off for females because they had larger litters and more pups were successfully weaned. Follow up experiments will determine if there are similar effects in human fertility and to determine exactly how the probiotics were improving overall health.

Prokaryotes and the Search for ET

I doubt we will find a technologically advanced extraterrestrial (ET) civilization similar to ours. However, the surprising ability of prokaryotes to survive in extreme environments on Earth has excited astronomers and the growing field of astrobiology. In 2015 NASA picked Europa, a moon of Jupiter with a liquid ocean under its icy surface as a destination for a new mission specifically to search for life. Later that same year, NASA confirmed liquid water on the surface of Mars, making it the second place in the solar system with liquid water on the surface. The finding is exciting because it improves the chances that life could be present elsewhere in the solar system.

Future missions to Mars, or any place in the solar system, with the intent to search for life will be tricky because we could easily contaminate any planet or moon with our own microbes. NASA has plans to send a manned mission to Mars in the 2030s. They will have to be careful when searching for life, especially considering, each person would carry a 100 trillion bacteria! However, the payoff could be incredibly huge if life was found elsewhere in the solar system. It would provide biologists an independent origin of life to study, shedding further light on the conditions needed for the origins of life from chemical evolution.

NASA is currently developing additional missions to search for exoplanets and signs of life. These include several large telescopes so sensitive they will be able to detect the atmospheric composition of exoplanets. One sign of potential life would be to find an oxygen-rich atmosphere, which would be a great indicator of photosynthesis taking place, a uniquely living process!

Endosymbiosis Gave Rise to Eukaryotes

The origin of eukaryotes is unique compared to the origins of other groups of organisms. Based on Darwin's theory of evolution, we know life evolves over time and modern species are the result of descent with modification as populations accumulate many small changes over time. But, when it comes to the evolutionary origin of eukaryotes, it's a bit more complicated. Prokaryotes did not slowly gain complexity in their cells, eventually evolving into the more complex eukaryotic cell. They have always been structurally simple and will remain that way.

Prokaryotes cannot evolve large complex cells only to haphazardly engulf some unsuspecting aerobically respiring bacteria. The main reason why prokaryotes will remain small comes down to energy limitations. Recall that prokaryotes make ATP by chemiosmosis across their cell membrane. If they were to grow larger, they would gain volume at a much larger rate than surface area, and it is surface area they need for generating ATP. Basically, as they become larger, ATP becomes increasingly scarce inside the cell, limiting the amount of energy available to them. Therefore, it took the rather unique event of two prokaryotes merging together to jump-start the evolution of eukaryotes, a non-Darwinian theory championed by Lynn Margulis in the 1960s. This radical new theory was called endosymbiosis, where 'endo' means 'within' and symbiosis means 'living together'.

When Margulis first proposed the endosymbiotic theory for the origins of eukaryotes, it was met with harsh skepticism. Part of the problem was that the technology required to support her theory wasn't available yet. Over the next few decades, support for the endosymbiotic origins of eukaryotes slowly continued to grow. By the 1980s, PCR was invented making it relatively simple to amplify small fragments of DNA, so they could be easily sequenced.

Sequencing DNA determines its primary structure, or the sequence of the four nucleotides. Recall that genetic information is stored in the sequence of nucleotides, just like letters of the alphabet convey information based on their sequences. Once PCR was available, mitochondrial DNA was sequenced revealing its distinct prokaryotic origins. Today, it is well established that the ancestor to mitochondria were once free-living bacteria using aerobic respiration.

With the theory of endosymbiosis firmly established, the next question centered on determining the identity of the host cell involved and how did the bacteria find its way into another cell? These are important questions, but with some basic knowledge about eukaryotic and prokaryotic cells, we can address the question of how it happened and who was involved. To begin, there are no prokaryotes with organelles or a nucleus. Whereas, modern day eukaryotes have a membrane-bound nucleus and a network of membrane-bound organelles used to compartmentalize the duties of the cell. Eukaryotes also have a complex cytoskeleton made of proteins that enable eukaryotes to move through their environment, maintain their shape, organize their organelles, divide, and engulf food or other cells from their environment.

Based on our knowledge of modern bacteria, it was unlikely that a prokaryote engulfed another prokaryote. It requires a lot of energy and a complex cytoskeleton to engulf a cell. So far, we have never found a prokaryote with a cytoskeleton capable of phagocytosis. Also, prokaryotes wouldn't produce enough ATP to carry out such an energy intensive feat. And lastly, we also don't see prokaryotes intermediate in complexity compared to eukaryotes. We do see some bacteria boring into other bacteria, and we have also found bacteria living inside of other bacteria. Based on this basic knowledge, we predict that the ancestor to mitochondria probably bored their way into the host cell rather than a primitive eukaryote engulfing another cell.

This brings us to the next question, who was the host cell? Based on 40 years of accumulated molecular data, the original host cell may have been an archaea. First, archaea and eukarya have ribosomes more similar to each other than either are to bacteria. Second, the DNA of archaea and eukaryotes are wrapped around small proteins called histones, whereas bacterial DNA is naked. With this information, a picture begins to emerge that eukaryotes formed when a bacteria bored its way into an archaea about 2 billion years ago. Additional evidence supports that the ancient archaea was potentially a methanogen, the same type of organism we met earlier in this chapter that makes methane from hydrogen gas and carbon dioxide.

That singular event led to the biggest reorganizations of cells in over 1.8 billion years! Endosymbiosis jump started the evolution of eukaryotes. Once this happened, Darwinian natural selection began to operate on those first eukaryotic cells, eventually giving rise to complex multicellular life forms including plants and animals. But first, let's explore a little eukaryotic diversity that doesn't include multicellular organisms such as plants and animals.

Eukaryotic Cells Are Structurally Complex

Eukaryotic cells, aided by the extra energy available to them, the benefit of multiple mitochondria producing lots of ATP, were able to make a giant evolutionary leap in their internal organization and size. Most eukaryotes are giant compared to the smaller prokaryotes, in some cases, thousands of prokaryotes could fit inside a single eukaryote. It takes energy to be so large, that's why mitochondria are present in almost every type of eukaryotic cell, including plants. Some eukaryotic cells may house upwards of a thousand mitochondria. I should note that a few highly evolved parasitic eukaryotes residing inside animals that over evolutionary time, their mitochondria became degenerate because they don't need a lot of energy to live in their host.

The DNA inside a nucleus distinguishes eukaryotic from prokaryotes. It is a double membrane similar to flattened sacs stitched together by proteins. In fact, eukaryotes get their name, new kernel, from having a defined nucleus. Therefore, most people consider the nucleus to be the defining character of eukaryotes. However, from an evolutionary point of view, it is really the mitochondria that define eukaryotes because they were present first and paved the way for the evolution of eukaryotes.

In addition to the nucleus, eukaryotes compartmentalize the functions of their cells with membrane bound structures called organelles. Organelles can be thought of as little organs similar to your organs that carry out different tasks in your body. Most organelles are involved in the manufacture and break down of molecules required for the cell's functioning. With organelles working together, eukaryotic cells can quickly respond to changes in their environment.

Since their origins nearly 2 billion years ago, eukaryotes have greatly diversified into plants, animals, and fungi. As we will learn in the next three chapters, eukaryotes are the only domain of life that evolved into true multicellular organisms, including plants and animals, with genetically identical cells carrying out different functions for the good of the colony. Many lineages of eukaryotes have also remained as single-celled organisms; some are quite important for life on Earth, while others have evolved into pathogens and parasites with elaborate life cycles.

The Single-Celled Protists

Protist is a generic term for any eukaryotic organism that is not truly multicellular, most are single-celled organisms. However, a few can form large colonies including, slime molds, kelp, and red algae. These large colonies are not true multicellular organisms because they lack distinct tissues found in plants, animals, or fungus. Despite this, there may be 200,000 species of protists in the world living in practically every environment, including our own guts alongside bacteria. Below, I introduce a few interesting types of protists.

Phytoplankton is Greek for "wandering plants", because they float in water unattached to any hard surface, moving with the wind and tides. They are not quite plants, but they do have chloroplasts, the energy converting organelle used for photosynthesis, making them autotrophs. Phytoplankton are incredibly important to the entire world. First, they form the bases of most marine and freshwater ecosystems. If you've ever eaten fish, especially from the ocean, the carbon in that fish was fixed by phytoplankton. Second, estimates show that phytoplankton produce upwards of 50% of the world's oxygen.

Although most phytoplankton are microscopic, too small to see with the naked eye, they are quite diverse in their shapes and sizes. Nowhere is this more evident in a group called the diatoms, perhaps the most abundant form of phytoplankton with an estimated 100,000 species. Diatoms show up everywhere; in rivers, lakes, marine ecosystems, and home aquariums. Diatoms are mostly single cells; however, some can form loose colonies.

Their unique cell walls are made of silicates and resemble two separate shells (valves). The tiny silicate shells of diatoms come in a wide range of shapes and sizes, some of them are among the strongest features in the world. It's been estimated that the strength of some diatom valves, if enlarged, could support the weight of an elephant. Studying the intricate patterns of the diatom valves provides engineers with new ways to build strong structures. Diatoms offer additional advances in biotechnology to produce technology at the nanoscale and to potentially help in the manufacture of solar panels.

Algae is a catch-all term for photosynthetic organisms including phytoplankton and cyanobacteria. There are thousands of types of algae living in practically every environment with water and sunlight, but they don't live in areas with extreme temperatures or pH. In recent years, there has been an interest in farming algae as a source of biofuel to replace fossil fuels. Algae fuel would have several advantages over fossil fuels. One is that it would reduce the input of carbon dioxide into the atmosphere, lessening our impact on global climate change. Second, it would be renewable source of energy. However, to farm enough algae to supply the nation's oil needs would require about 1/7 of the area we use to currently grow corn. That's an area about half the size of Maine. The current obstacles to algae-based oil include engineering algae to cheaply produce oil and building the infrastructure to grow and harvest the algae-made oil.

The single-celled *Euglena* has characteristics of both plants and animals. Well almost, it's not a multicellular organism, so in that way it is neither a plant nor an animal. They do have chloroplasts for photosynthesis. However, when there is not enough light to power photosynthesis, they switch to becoming heterotrophs. Specifically, they engulf other species of algae. As you can guess, their heterotrophic and autotrophic tendencies made it difficult to classify. Fortunately, modern DNA sequencing has shed light on their diversity and classification. We have identified at least 800 species, most found in most freshwater, and in some cases in large enough numbers to turn the water green.

Sometimes, the lines between multicellular organisms and single-celled organisms becomes blurred, especially when the organism in question is the size of a golf ball. I'm talking about slime molds, an informal name given to a type of protists that aggregates into large groups to reproduce. Most of the time, a slime mold is basically just a bunch of individual cells hanging out by themselves. Then some environmental signal, such as a change in food availability, causes them to come together into a giant amorphous body. Once they come together, they detect odors in the air and move towards it. Even though a slime mold is made of individual cells, it can behave similar to a multicellular organism.

Sponges represent another example of the blurred line between colonial cells and animals. Animals almost certainly evolved from a single common ancestor about 600 million years ago and are clearly related to sponges, but should we classify sponges as true animals? Most biologists classify sponges as an animal because they have specialized cells. However, they lack true tissue layers, muscles, nervous system, internal organs, symmetry, and specialized genes used for laying out a body plan.

The clear connection between sponges to protists and animals is a specialized collar cells used to move water through the sponge. Collar cells have a flagella surrounded by a small collar, hence the name. Similarly, a protist called a *Choanoflagellate* looks almost identical to the collar cells of the sponges. Most of the time, choanoflagellates are single-celled, but if they sense a predator in the area, they form a small colony to avoid being eaten. Perhaps the origins of animals began when the ancestor to choanoflagellates began to form colonies to avoid predation.

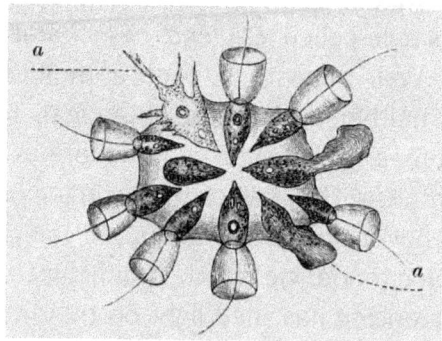

On the left, a choanoflagellate, perhaps the ancestor to all animals. They resemble the collar cells in sponges and several species form loose colonies. Their evolutionary relationship to animals was proposed in the mid 1800s. Recent evidence from DNA supports the relationship.

Chapter 8

Plant and Fungal Diversity
and
The Greening of the Earth

For the first four billion years of Earth's history
The land was barren, devoid of plant and animal life;
Then plants colonized this barren landscape creating the
first forests resulting in the greening of the Earth.

Introduction

For the first 4 billion years of the Earth's history, the surface of the Earth was barren; no forests and no animals were to be found anywhere. Then, about 470 million years ago, the first plants and fungus began to colonize the land, most likely together. However, the story of plants begins about a billion years ago when the eukaryotic ancestor to plants acquired their photosynthetic ability by engulfing a cyanobacteria. These small photosynthetic bacteria eventually evolved into chloroplasts, the organelle where photosynthesis takes place in modern plant cells.

The story of terrestrial plants began as small mosses evolved from algae about 470 million years ago. Since those first mosses along stream banks, plants grew taller by evolving lignified cell walls, leaves, roots, stomata, and a vascular system. The result was that it transformed the surface of the planet from a barren wasteland to lush forests teaming with life. Another evolutionary milestone occurred in the early Cretaceous about 120 million years ago with the appearance of flowering plants. Today there are over 300,000 species of flowering plants alone, which would take many volumes of books to cover their amazing diversity.

Fungi most likely colonized the land at the same time as plants about 470 million years ago. They too are multicellular, but are heterotrophic, a quality making them a vital component for the world's ecosystems as decomposers of dead plant and animal material. By producing a variety of enzymes, fungi break down organic matter and even speed up the erosion of rocks, releasing their minerals and making them available to life. In addition to their roles as decomposers, some fungi are pathogens and implicated in amphibian declines, athlete's foot, crop destruction, and witch hunts. Not to mention that fungi are used to brew beer, bake bread, make antibiotics, while others can cause hallucinations and death! Although fungi have evolved into many species, their total diversity is not well known because many are microscopic.

Part I: Plants
The Origins of Plants

Although plants are photosynthetic, I want to point out once again that prokaryotes evolved photosynthesis. Recall from Chapter 3 that photosynthesis is a series of chemical reactions where the relatively inert gas carbon dioxide (CO_2) is "fixed" into organic molecules, and free oxygen (O_2) is released as a byproduct. To put it another way, photosynthetic organisms make their own food from carbon dioxide and water by using the energy in sunlight.

The origin of plants dates to about a billion years ago when an early eukaryote engulfed a cyanobacteria. Eventually the cyanobacteria would evolve into chloroplasts inside the cells and is a second example of endosymbiosis in eukaryotes. Today, chloroplasts are the organelles responsible for photosynthesis in plants and different types of eukaryotic algae. Algae is a general term for different groups of eukaryotes from single-celled forms to macroalgae, such as kelp, that blur the lines between colonies of individuals and true multicellular organisms.

About a billion years ago, algae had become vital components of early marine ecosystems, eventually providing additional resources for the first animals 542 million years ago. However, it took more than 500 million years from the time a eukaryote engulfed a cyanobacteria to the evolution of the first plants able to live on land. The transition to living on land presented several major hurtles. First, you had to worry about drying out, so ways of preventing water loss had to evolve. If you live in water, then gravity is not much of a problem, but if you're on land, then you hmustreinforce your cells walls to support yourself. If you're in water, it's easy to absorb all the nutrients you need directly into your cells from the water. But if you live on land, you can only absorb nutrients and water from the soil. And lastly, you can only absorb nutrients from the ground, so you need ways of moving material between the leaves and roots of plants.

Despite these obstacles, after four billion years of the Earth's surface being devoid of vegetation, the first plants colonized the land perhaps some 470 million years ago in the Ordovician. At first, plants were small and lived near water or moist environments along streams and lakes. Based on the fossil record, molecular data, and common features of all modern plants, they most likely evolved from a single algae ancestor that strongly resembles modern day *Chara*, a resident of freshwater lakes and streams.

What Makes a Plant Unique?

Before getting started, I want to point out that at the cellular level all eukaryotes, including animal, plants and algae remain fairly similar in how they function. They possess mitochondria for aerobic respiration. Meaning they use oxygen to efficiently produce ATP just like animals. Their DNA is in the cell nucleus and tiny membrane-bound organelles compartmentalize the functions of the cell. Although though plants have mitochondria and use oxygen, their rate of photosynthesis, during the day at least, is higher than their rate of oxygen consumption from respiration. At night, plants continue to use oxygen and give off carbon dioxide, just like an animal.

Plants are autotrophs, they use the energy in sunlight to fix carbon dioxide to make all the organic molecules they require. Although, in biology, you can always find exceptions to any rule, a few parasitic plants have lost their photosynthetic ability. Because most plants make their own food, they form the base of terrestrial ecosystems, meaning lots of animals eat plants. As I'll show later, they are not without their defenses! Plants have cell walls made of cellulose to maintain shape and rigidity. To further reinforce their cell walls so they can grow tall, plants use strong molecules called lignin, making the cell walls even more rigid.

Unique tissues and organs separate plants from algae. The most obvious organs are the leaves, an adaptation that increases surface area for photosynthesis. To prevent water loss from the leaves, they have a waxy upper surface and tiny openings on the underside called stomata open or close depending water availability. Roots anchor plants to the ground and absorb nutrients and water. A vascular system connects the roots to the leaves in plants.

Another defining feature of true plants is a unique life cycle where they alternate between multicellular haploid and diploid forms, in what is a called an alternation of generations. To understand alternation of generation in plants, it can be helpful to review our life cycle. Recall that humans make single-celled haploid gametes (sperm and egg) each carrying a single copy of a genome on 23 chromosomes. The sperm and egg fuse to form a zygote, which grows into a new diploid organism carrying two copies of the genome with 23 pairs of chromosomes, or 46 chromosomes total. Our adult form is diploid and multicellular; however, we produce single-celled gametes that do not divide or grow into a multicellular organism.

Plants are different, they have a multicellular haploid form called the gametophyte (*phyte* means plantlike), which produces single-celled gametes that fuse to form the diploid sporophyte. When you look at a tree, or the grass in your lawn, those plants are the sporophytes. In most modern plants, the gametophyte has become almost microscopic. It is this alternation of generations with plants having two multicellular life-stages, the sporophyte and the gametophyte that makes plants a unique lineage of organisms, separating them from algae. Although in flowering plants, the gametophyte is very small, consisting of just a few cells.

The Evolution of Plants

It's not quite clear when the first plants colonized the land. Fossil records support a date of approximately 450 million years ago, with some evidence pushing that date back to 470 million years ago. These ancient plants strongly resembled modern day mosses known as bryophytes, small plants found living in moist environments. Similar to modern mosses, they lacked true leaves, roots, and a vascular system to move water and nutrients from one end of the plant to the other. They also lacked ways to prevent drying out in the atmosphere, restricting them to moist areas. Although these early mosses had cell walls made of cellulose, they lacked lignin, so they were unable to grow very tall.

For nearly 30 to 50 million years, plants remained relatively small and limited to growing in wet or moist areas. By 420 million years ago, the first vascular plants had evolved resembling modern-day ferns. Evolving a vascular system for transporting water and nutrients from the roots to distant parts of the plant was a major evolutionary break-through that when combined with lignified cell walls and true roots, allowed plants to grow tall. With these new features, the first forests began to emerge and were well underway by 400 million years ago. For the first time in 4.2 billion years of Earth's history, the surface was green and teeming with life. The spreading of plants on land also greatly reduced erosion, increasing the amount of soil, further helping the spread of plants and expanding terrestrial ecosystems.

From 360 million years ago to about 300 million years ago, the climate remained wet and warm. Carbon dioxide levels were approximately 2,000 ppm, more than 5 times higher than today, keeping the Earth rather warm. The continents had drifted near the equator, and large tropical swampy forests dominated by giant fern trees covered much of the land surface. Paleontologists refer to this time between 359 – 299 million years ago as the Carboniferous Period because large coal deposits were formed during this time. The wet, swampy conditions were conducive to coal formation by preventing oxygen from breaking down the organic matter and releasing it back to the atmosphere as carbon dioxide.

Over millions of years, the accumulation of coal deposits reduced carbon dioxide levels in the atmosphere causing the climate to become cooler and drier. By the end of the Carboniferous Period, the Earth entered an ice age as carbon dioxide levels dropped to about 250 ppm. In Chapter 14, I will go into more details about how life has influenced the atmosphere and climate. As the Earth's climate cooled 300 million years ago, the first seed plants known as gymnosperms evolved. Gymnosperm means 'naked seed' because the seeds develop on the surface of the leaves, scales, or cones. The protective coverings around the seeds and pollen preventing water loss made them well adapted to the cooler drier environments. By 251 million years ago at the end of the Permian Period, gymnosperms dominated the Earth's surface. Even today, they form the bulk of our northern forest, still dominating in cooler and drier environments.

During the late Mesozoic Era, a time when dinosaurs were in their heyday and the first birds were taking flight, plants evolved flowers. The arrival of flowering plants had an enormous impact on the future of diversity. Flowers were a novel adaptation to attract pollinators to spread their pollen. Conifers use the wind to spread their pollen; because the process is random, they must produce copious amounts of pollen that they release to the world around them. If you've ever lived near pine trees, you may be familiar with the yellow pollen that covers your car in the spring. Flowering plants produce less pollen, instead they invest more of their energy into flowers and nectar, a sugary snack to entice animals to visiting the plants. Many flowering plants also evolved elaborate fruits to aid in seed dispersal.

Modern Plants Are Very Diverse

There are about 300,000 species of plants in the world, including bryophytes, ferns and their allies, conifers, and flowering plants. Modern plants range in size from little more than a tiny leaf floating on water, to giant sequoia trees growing over three hundred feet tall. They are found on every continent, including Antarctica, and live in extreme environments from the very cold, to hot and dry deserts, some have even returned to water. A few species of plants in nutrient poor areas turned to carnivory, while others became parasitic giving up on photosynthesis all together. I'll start introducing plant diversity with the bryophytes.

Even today, small bryophytes, relics of the first plants from a time long gone, are found on every continent, including Antarctica. They are called non-vascular plants because they lack a vascular system and they also lack leaves and roots. Surprisingly, there are more than 16,000 species of bryophytes living today. Bryophytes are actually a generic term for three distinct groups of non-vascular plants called hornworts, liverworts, and the true bryophytes.

Most of the seedless vascular plants have gone extinct, but one group called the horsetails or *Equisetum*, can still be found growing along stream banks and other wet areas. They date back to the Devonian some 400 million years ago when they were the dominant understory vegetation.

Ferns are also an ancient lineage of seedless vascular plants dating back to about 360 million years ago. Today, about 10,000 species of ferns are found on most continents, except Antarctica. Most are understory plants, but a few species are trees. They do not produce flowers, seeds or pollen, instead they produce tiny spores which develop in specialized structures on the underside of their leaves.

Gymnosperms were the first seed plants and producers of pollen, evolutionary adaptations to cooler and drier environments. In seed plants, a tough coat forming the pollen protects the male gamete. Gymnosperms lack flowers and rely on the wind for pollination, so they produce lots of pollen. Today, about 1080 known species of gymnosperms inhabit the world. The most common gymnosperm are the conifers, which produce seeds in their cones. Conifers include common plants like, pines, junipers, redwoods, spruces, firs, and cypress trees. They form the largest forests in the world, stretching across Europe, Asia, and North America dominated by cold hardy conifers, living in the harsh environments.

Modern conifers hold records for the oldest and largest organisms on the planet. A candidate for the oldest living organism on the planet is a bristlecone pine estimated to be 4800 years old. It's difficult to determine the largest single living organism on the planet, but it may be a Sequoia in California nicknamed General Sherman. It is larger than 24 blue whales, or 40,000 people. That's more people than are found in most universities in the U.S. One of the most interesting Gymnosperms is a plant called *Welwitschia* found only in the Namibian desert in Africa. It has only two leaves that grow continuously throughout its life, which some estimates place over 2,000 years!

The evolution of flowering plants, or Angiosperms, represents one of the greatest success stories in life, showing how one evolutionary adaptation creates new opportunities for additional adaptations. The arrival of flowering plants was a driving force in the diversification of animals over the last 100 million years, as diversity gives rise to diversity. Today, with over 300,000 species, flowering plants dominate plant diversity. The greatest diversity is in the tropics, compared to just over 1080 species of Gymnosperms, a group of plants that has been around twice as long as flowering plants.

Most flowering plants are placed into one of two large groups based on the number of leaves an embryonic plant produces. The monocots make one embryonic leaf and eudicots make two embryonic leaves. Another way to easily distinguish the two groups is by observing the patterns of veins in the leaves. Monocots include plants with parallel veins in their leaves. Examples are grasses, including bamboo, corn, wheat, and rye. Some monocots have beautiful flowers such as orchids and lilies. Monocots are typically not woody, lacking the ability to grow into large trees. The eudicots are easily recognized by the branching pattern of veins in their leaves and include broad-leaf trees, bushes, and many familiar wildflowers with including, roses, peas, and sunflowers.

Today, flowering plants are found on every continent, including small grasses living in Antarctica. In the tropics and throughout much of the middle latitudes, broad-leaf trees dominate the forests. Broad-leaf trees in the tropics keep their leaves year-round to take advantage of the abundant sunshine. Although, they will replace them throughout the year. In temperate regions, where winter temperatures dip below freezing, many broad-leaf trees are deciduous, dropping their leaves in the fall, only to regrow them in the spring once temperatures warm.

In the American deserts, cactus evolved several features to survive the arid conditions. Their leaves have been reduced to sharp spines providing protection to the plant while at the same time preventing water loss. As a result, cactus leaves are not photosynthetic, but the bark is, giving cactus their overall green appearance. Cactus are not the only drought tolerant species. In Africa, a genus of plants called *Euphorbia* also evolved reduced leaves to prevent water loss and larger stems capable of photosynthesis and holding water.

In contrast to cacti, some flowering plants returned to aquatic and marine environments. Two examples are turtle grass (*Thalassia testudinum*), a common plant in the Gulf of Mexico and eel grass (*Vallisneria americana*) in eastern streams, both resemble terrestrial grasses but are unrelated. Their similar form is an example of convergent evolution when two species evolve to look similar because natural selection was operating in similar environmental conditions. In this case, a flat grass-like leaf is beneficial for maximizing surface area while reducing drag in the water.

Within the Cenozoic Era of the last 65.5 million years, the grasses evolved to become one of the dominant terrestrial plants. Today, about 12,000 species of grasses cover nearly 20% of the terrestrial surface. You can find grasses living in parts of Antarctica, wetlands, forests, and tundra, not to mention that grasses dominate tropical savannas and temperate grasslands. Grasses are also one of the most important plants for humans and account for much of our food production. Some of the most important grasses for agriculture include sorghum, rice, wheat, and corn. Rice alone may account for 1/5 of all the calories consumed by humans.

Modern flowering plants have many diverse lifestyles and some plants have evolved "alternative" lifestyles. Mistletoe, the green plant seen growing on the limbs deciduous trees in the winter, is actually a parasitic plant that sends its roots into the branches of other trees robbing them of nutrients. However, many species of mistletoe retain their photosynthetic abilities, any green plant is photosynthetic. Not only is mistletoe a parasite, it's quite poisonous producing a chemical called phoratoxin, and eating it causes nausea, diarrhea, and blurred vision. But, let's not forget that it has medicinal properties as well, including potential cures for cancer

Growing on the floor in deciduous forest of the eastern US are small white plants called Indian pipes (*Monotropa uniflora*). They are a parasitic plant tapping into the roots of beech trees (*Fagus grandiflora*), allowing them to grow in the dark understory of forests where there isn't enough light for photosynthesis. Lacking in the green pigment chlorophyll, they are white and therefore have totally lost the ability for photosynthesis.

Some flowering plant species that live in very nutrient poor soils are carnivorous. They evolved modified leaves to attract and catch unsuspecting animals making a meal out of them. Perhaps the best-known example is the Venus flytrap (*Dionaea muscipula*), a resident of swamps in North and South Carolina. Insects are attracted to the bright red inside of the highly modified leaves that rapidly close. If an insect lands on the leaf and bends two or more of the "trigger hairs" the two leaves rapidly close catching the insect. Additional movement by the prey causes the leaf to seal off the insect and release digestive enzymes.

Pitcher plants also use modified leaves to catch insects. In this case, their leaves are shaped into a pitcher capable of holding water. To help trap the insects, the inside of the leaves forming the pitcher have small hairs pointing downward making it difficult for insects to land on the sides. Eventually, the exhausted animal falls into the water where it is slowly digested, and its nutrients are absorbed by the plant.

Found alongside pitcher plants are sundew (*Drosera* sp.) plants. Their modified leaves exude a sticky sap that catches small flying insects. Once an insect is caught, the leaf rolls up to digest the animal. Often seen floating in ponds are bladderworts (*Utricularia* sp.), a plant with the most complex traps among plants. They use modified leaves that form a bladder underwater and then actively pump the water out forming a vacuum. When a small animal touches the "trigger hairs", the trap quickly opens and sucks in the prey. Once inside the bladder, the plant releases enzymes to digest the prey.

In addition to flowers for attracting pollinators, flowering plants evolved fruits, a seed-bearing structure formed from parts of the flower. Fruits aid in seed dispersal, and like flowers, they too have become specialized in their modes of dispersal. Fruits come in a very wide variety, ranging from raspberries, to apples, oranges to large coconuts. Technically, many "vegetables" we eat including nuts, corn, wheat grains, tomatoes, chili peppers, rice, and beans are actually fruits.

Some plant species coevolved to form symbiotic relationships with animals where the fruit provides nutrition to the animal and, in turn, the seeds are dispersed from the parent. If you've seen more plants growing along a fence-row, it is from birds eating fruits and dropping seeds when perched on the fence. In fact, the seeds of raspberries and blackberries won't germinate unless they pass through the gut of a bird. If you've ever picked up a sandspur or cocklebur, those are seeds that evolved to get caught on mammal fur for dispersal. Some seeds such as dandelions and maple seeds are small and dispersed by the wind.

The Connection Between Plant and Animal Diversity

Plant and animal diversity are closely tied to each other. There are an estimated 300,000 species of flowering plants, the end result of 120 million years of evolution with animals. The rate of diversification for the past 542 million years has not been constant, but has undergone several spurts of rapid diversification, including the current period of high diversity because of flowering plants. Plants not only provide nutrients and energy to animals, but plants also provide structure to ecosystems, creating additional habitats for animals. In addition to creating structure, animal diversity has paralleled the diversity of flowering plants as both flowers along with their fruits and seeds have provided additional opportunities for diversification.

Every flower has an evolutionary history; it is the result of millions of years of natural selection that has honed its ability to efficiently spread its pollen, often by attracting a specific pollinator. Sometimes, flowers become incredibly specialized as they evolved to attract only one particular species of pollinator, while other flowers evolved to attract as many species as possible. The end result is that flowering plants diversified into a myriad of colors, shapes, and sizes. For example, large white flowers opening at night attract bats and nocturnal moths. Bright yellow sunflowers attract a wide variety of insects including bees, butterflies, and flies. Additionally, their seeds attract birds to aid in their dispersal away from the parent.

Many flowers are quite aromatic, producing pleasant smelling odors to attract many types of insects. Even humans enjoy the sweet smell of roses. However, not all flowers have a pleasant smell like sweet roses. In southeastern Asia, is the world's largest flower, at least by weight. Except you wouldn't want to smell it because it smells like rotting flesh. Commonly called the "corpse flower", its stinking flower attracts flies to serve as its pollinators. This large flower in the genus *Rafflesia* is a parasite of a particular vine, so there isn't much of a plant to look at when it is not blooming. Ironically, the largest flower by size is the corpse lily *Amorphophallus titanum* of Sumatra. It is very rare, but its flower can grow up to 12 feet tall and weigh about 200 pounds.

In my opinion, the orchids grow some of the most beautiful flowers in the world. In many cases, a species of orchid has become so specialized, it can only be pollinated by one specific species of insect or bird. This happens through coevolution when two or more species mutually interact over time driving the other's evolution. A classic example of coevolution is the hammer orchid (*Drakaea glyptodon*) that has evolved to be pollinated only by one species of wasp. To attract a male wasp, the flower resembles the female, but it also emits a pheromone that mimics the female. The male wasp will land on the flower and attempt a copulation, which will invariably fail. Fortunately for the plant, the wasp gets a full dose of pollen to spread to the next flower. There are about two thousand species of orchids including some interesting and beautiful species such as the Monkey Face Orchid, Moth Orchid, Bee Orchid, Flying Duck Orchid, Tiger Face Moon Orchid, Dove Orchid, and White Fringe Orchid.

Coevolution is not limited to flowers and their pollinators. Another example of coevolution between plants and animals takes place with the acacia ants (*Pseudomerymex*) living on bullhorn acacia trees (*Acacia cornigera*) throughout Central America. The tree grows large hollow thorns where the ants live, and the leaves make a sugary drink supplied to the ants. In return, the ants fend off any herbivores that try to eat the plant.

Tropical rainforests are the heart of plant diversity due to rapid rates of evolution driven by competition, symbiotic relationships, or for other reasons not yet known. In the Amazon Rain forest, there may be upwards of 300 species of trees per acre, and approximately 1000 flowering plant species. In addition to their diversity, trees and other plants also add structure to communities creating additional habitat that can be exploited by animals, fungi, protist, and bacteria. In fact, most of the diversity in rain forests is in the canopy.

The forest canopy includes the upper parts of all the trees like the trunk, stems, and leaves. The branches of large trees become covered themselves in plants, including bromeliads and orchids, all providing additional resources numerous animals. Some estimates suggest that an acre of lowland Amazonian rain forest contains upwards of 30 billion arthropods. Another study in Ecuador discovered over 100,000 species of plants and animals per acre in the canopy alone.

Plants are Chemical Factories

Plants are often passed over as boring compared to the more charismatic animals. But, we shouldn't underestimate plants, it turns out that plants are not as passive or defenseless as once thought. They make up for their sessile lifestyle by making thousands, if not millions of different chemicals for various purposes including defense against herbivory, communication with other trees, and attracting pollinators to aid pollination. The sweet smell of roses, the savory taste of garlic, and the heat of chili peppers are all sensations caused by chemicals created by plants. Sometimes they are referred to as phytochemicals, because they are made by plants. Let's take a closer look at some of these chemicals that are both beneficial and harmful.

If you've ever improved the taste of your favorite foods by adding herbs and spices including, rosemary, thyme, garlic, cinnamon, oregano, pepper, *etc.*, then you know that plants produce many aromas and taste that we enjoy. Although, we enjoy flavorful food, plants don't always make them for us, many of the chemicals found in our herbs and spices are used by plants to fend off insect herbivores. What is harmful or taste bad to an insect may actually be beneficial to humans. Indeed, scientific research backs this claim. In one large study, scientists showed that people who cooked with a variety of herbs and spices had lower rates of disease and parasite infections.

Two of the most common drugs used include caffeine and nicotine, two examples of compounds made by plants to prevent herbivory from insects. In small amounts, they are both stimulants and, unfortunately, addictive, although caffeine not as much as nicotine. Scientific studies have shown that small amounts of caffeine from tea and coffee improves cognitive function and reaction times. However, too much caffeine can cause health problems including rapid heart rate, loss of sleep, and anxiety in some individuals. Ironically, insects respond similarly to caffeine, it wakes them up, or makes them a bit more twitchy. Unlike us, they don't get pleasure from caffeine, it's just annoying to them, so they avoid it.

Living in New Mexico, most of us enjoy adding green chili peppers to just about everything from pizza to breakfast burritos to cheeseburgers.

In fact, I would consider green chili to be the plant equivalent to bacon for its versatility and ability to improve the flavor of so many foods. Some industrious cooks go so far as to roast chili peppers stuffed *and* wrapped in bacon! The active ingredient in chili peppers making them hot is called capsaicin, an irritant that we perceive as heat. Contrary to popular belief, the seeds of peppers lack capsaicin, but the areas around the seeds do. Birds spread pepper seeds because the seeds pass through the digestive tract unharmed. Mammals, on the other hand, use teeth to grind the peppers, destroying the seeds. Therefore, it's most likely that peppers evolved capsaicin to prevent mammals from eating the peppers because birds are unaffected by capsaicin.

Although, capsaicin is an irritant, it can be applied to the skin to relieve minor muscle aches and pains. Claims have been made that it may prevent cancer; however, we lack evidence to support this claim, but that also means we should continue to study its benefits to our health. A very large study did show that eating more spicy foods was correlated with lower death rate, but the reasons for the lower death rate are not known. A limitation of these studies is that other factors could have led to lower death rates. Perhaps people who eat spicy food walk more, in this case, walking most likely contributed to longer life.

Over 50% of our medicines come from plants. For thousands of years, people have turned to plants for remedies to cure all kinds of ailments including nausea, toothaches, diarrhea, aches and pains. Perhaps one of the most widely used medicines from plants is called Acetylsalicylic acid, better known as the active ingredient in aspirin. This chemical was first isolated in the 1850s from the bark of the common willow tree (*Salix*), although the therapeutic properties of willow trees have been known for over 2400 years, as the Greek philosopher Hippocrates prescribed it for headaches. In California, grows the endangered Pacific yew tree (*Taxus brevifolia*) whose bark contains an anticancer chemical called paclitaxel. It works by preventing cells from dividing. Researchers at Florida State University learned how to make it in the lab so that it could be used as an effective cancer treatment, without having to harvest the tree.

Primates, including humans, don't make vitamin C, instead we get it from the foods we eat. In recent years some plants have earned the label as a super food because they are packed with vitamins and minerals needed to keep us healthy. While there is no consensus on the exact ranking of foods based on their nutrition, there are some good guidelines. Green leafy vegetables including kale, spinach, and broccoli are considered to be very healthy and come recommended as part of a healthy diet. Other healthy foods include blueberries, garlic, and peppers. The idea is to eat a varied diet with multiple colors of fruits and vegetables to ensure good health. For example, red bell peppers actually contain more Vitamin C than oranges and other citrus fruits.

While many plants are beneficial to humans, some are quite poisonous. A poison is any substance that causes harm to an organism and a toxin is more narrowly defined as a poison made by an organism. Poisonous plants and animals make toxins, this means that harmful substances such as heavy metals, commercial insecticides, and plastics are poisons, but they are not toxins. Plants produce their toxins to prevent disease, fungal infections, herbivory from mammals and insects, and to prevent competition from other plants. Perhaps the most infamous poisonous plant is Hemlock in the genus *Conium*. In 399 BC, the Greek Philosopher Socrates was sentenced to death; to carry out the sentence, he drank a cup of hemlock tea and died within a half hour.

The deadliest plant in the world is *Nicotiana rustica*, commonly known as tobacco. It makes the powerful stimulant nicotine an incredibly addictive opiate related to morphine and caffeine. Every year, 5-6 million people die of preventable deaths, the direct result of smoking tobacco. No other plant causes more death than tobacco.

When attacked, plants quickly deploy their chemical defenses to ward off herbivory. However, in the 1980s, a surprising discovery found that plants actually talk to each other. Rather than using sounds waves like animals, they release chemical signals warning nearby plants of danger from an herbivore. When a plant is getting grazed by an insect or mammal, it begins to produce compounds that will make it taste bitter, so the animal moves to the next plant. But, the plant has also sent a chemical message through the air to other plants. Upon receiving the signal, the nearby plants begin to produce chemicals making their leaves bitter to the grazing animals.

Part 2: Fungi
Origins of Fungi

The origin of fungi has been difficult to determine because they don't fossilize well. To further complicate the story, the fossils we do have can be difficult to interpret, and DNA evidence has been varied. However, by piecing together all the available evidence, the origins of fungi began somewhere between 700 million and 1.5 billion years ago when their ancestors split from other eukaryotes. The last common ancestor of fungi and animals was probably unicellular, and multicellularity evolved independently from the animals. In fact, some fungi, including baker's yeast have remained unicellular.

Similar to plants and animals, fungi originated in water and colonized the Earth's surface later. However, fungi may have been among the first organisms along with plants to colonize the land around 450 million years ago. Evidence from the fossil record indicates that fungi quickly formed symbiotic relationships with the earliest plants, helping them to survive on land by improving their ability to get nutrients from the primitive soils. By the Devonian 400 million years ago, fungi were common in the fossil record and were an integral part of terrestrial ecosystems. Most of the modern groups of fungi were present by the end of the Carboniferous 299 million years ago. The first mushroom-forming fungi were present by the mid-Cretaceous, about 90 million years ago.

Fungi Exhibit Diverse Lifestyles

Fungi, like plants and animals, are made of eukaryotic cells. Independently of plants, fungi evolved a cell wall made of chitin, which is even more rigid than cellulose. Historically, most generally presumed that fungi were more closely related to plants than to animals because of their superficially plant-like characteristics including a cell wall, sessile life style, root-like structures, and plant-like growth. However, molecular data and the fact that they are both multicellular heterotrophs lacking chloroplasts support the closer relationship to animals and not plants. Heterotrophic organisms must acquire energy by breaking down organic molecules they obtain from their environment. Mushrooms accomplish this by using root-like structures called hyphae that produce enzymes to break down material so they can absorb it.

With a current estimate of 100,000 species, we still don't know the global diversity of fungi. Although, estimates range as high as one million species. One reason why fungus are so diverse is easily described with the simple statement: Diversity gives rise to diversity. As we saw with the evolution and subsequent diversification of flowering plants, insects also quickly diversified. Fungi followed the same pattern as plants, they have greatly diversified mirroring plant and animal diversity. New species of fungi emerge as they co-evolve with plants and animals. In fact, whenever you observe a unique looking fungus, it may only be found on a certain type of tree, root, cheese, or animal.

We can broadly divide fungi into two groups; macroscopic fungi that are easy to see, and microscopic fungi that require a microscopic to be observed. Most familiar fungi are macroscopic, including mushrooms, lichens, and various molds. However, most fungi remain inconspicuous due to their small size and cryptic nature. Unfortunately, many types of microscopic fungi cause diseases and discomfort, like the itchiness from athlete's foot.

More than once I've cultivated some pesky mold that seemed to appear out of nowhere on my leftover lunch that stayed too long in our refrigerator. Although, I get sad having lost my leftovers to some fuzzy looking mold, but I can't help wonder how it got there? What type of mold is it? Is it unique to bread or the cheese, or the fruits? And how does it make a living on my left-over lunch?

Common on bread, black bread mold is usually caused by the species *Rhizopus stonlifera*. Because fungi are heterotrophic, the hyphae release chemicals into their environment to break down and absorb organic molecules. Fungi also wage chemical warfare by producing chemicals to protect themselves from bacterial infections and to prevent competition from other molds growing near them. Unfortunately, if your bread becomes moldy, it's best to not eat it because many of these chemicals are toxic to us and the hyphae may be growing throughout your bread.

Before we all become mold-phobic, don't forget that molds make some of our favorite cheeses. Blue cheese is made by adding a mold in the genus *Penicillium* to the final product providing the cheese with its distinctive smell and blueish color. If you recognize the name of that mold, it is because our first wide-spread antibiotic used to treat bacterial infections was isolated from a compound made by a similar species of mold! Pathogenic bacteria harm plants, animals, and fungi. As a result, molds in the genus *Penicillium* protect themselves from bacterial infections by making antibiotics that prevent bacteria from forming a cell wall, killing the bacteria.

When walking through a forest, on campus or, even in your backyard, you can see mushrooms popping up in circles or in lines, especially after rain. Mushrooms are a small part of the whole organism used for reproduction by producing tens of thousands, and in some cases, billions of spores to be spread by the wind. The majority of the organisms lives below ground and made mostly of hyphae. In some cases, the total below ground mass of some fungi are thought to be massive, and may be among the largest organisms on the planet. One example is found in the Blue Mountains of Oregon where a honey fungus (*Armillaria ostoyae*) is thought to be 2.4 miles across and between 1900 and 8650 years old. Because most of the fungus lies underground, it makes it hard to know whether it is truly a single organism. However, all the individuals are genetically identical and recognize each other just like the cells in our own body recognize each other.

As a decomposer, fungi play a vital role in our ecosystems by breaking down organic matter so that the nutrients and minerals can be recycled and used again. Without the action of decomposers, much of the world's nutrients would soon be locked up in dead plants and animals.

Mutualists

Fungi maybe well known as the decomposers of ecosystems, but many types also form mutualistic relationships, mostly with plants and algae, and in some cases, they are even farmed by ants! *Mycorrhizal* fungi form close relationships with the roots of many plants species improving the growth of the plant. This makes sense because fungi produce powerful enzymes capable of break down organic material, or even rocks, liberating essential nutrients and minerals making them easier for plants to absorb through their roots. Some fungi form such a close relationship with plants that their hyphae even enter and live within the root's cells.

If you've ever been hiking in the woods, you've probably seen lichens growing on the trees, on the rocks, or even on the ground. Lichens are incredibly hardy due in part to their mutually beneficial relationships with algae. To benefit the algae, the lichens obtain minerals required for the algae to grow. The tough cell wall of the lichens prevents desiccation while other chemicals make them toxic preventing herbivory. The algae benefit the lichens, by providing them with a source of organic molecules and energy.

Humans are not the first farmers. That distinction goes to ants, those venerable insects that appear in just about every terrestrial ecosystem. Although, not every species of ants are farmers, this title goes to the leafcutter ants of Central and South America. In all, 47 species of these ants harvest plant material to grow a type of fungus called *Mycelium*. Next to humans, leaf cutter ants exhibit the most complex societies in the animal kingdom. Colonies can reach upwards of 8 million individuals and cover several thousand square feet.

The relationship between the ants and their fungi also involves a third player. Growing on the ants is a species of bacteria that secretes antimicrobial compounds to prevent undesirable bacteria from causing problems for the fungi. Recall that many types of plants produce toxic chemicals to prevent herbivory. If a leaf turns out to be toxic to the fungus, it will send out chemical signals to the ants communicating to them not to collect that particular type of leaf.

Parasites

Can a fungus really be a parasite? The answer to this is: yes, but it's complicated. Some parasites form close symbiotic relationships with their host. Rather than being mutualistic where both partners benefit from the arrangement, parasites benefit at the expense of their host. As we will learn in the next chapter, many parasites have complex life cycles and some alter the behavior of their animal host to improve their chances of transmission. The same is true of the *Cordyceps* fungus, which turns many insects into a living zombie.

When the *Cordyceps* fungus infects an insect, it will eventually kill the insect. Once an insect becomes infected with the fungus, the fungus alters the insect's behavior by causing it to move to an open area and latch on to a branch or leaf only to die in place. Once the unlucky insect dies, the fungus begins to act more fungus-like, slowly decomposing the animal to acquire nutrients. After a short time, it grows out of the insect and release its spores to infect other insects. There are about 400 known species of *Cordyceps* fungi, each one infecting a different species of insect. As you can see, the *Cordyceps* fungus acts like a parasite because it infects a live organism and alters its behavior to improve its chances of infecting another insect. Some observations have shown that ants infected with the fungus will go so far as to climb a branch above their own colony. If an infected ant is spotted by another individual ant from that colony, the infected ant will immediately be carried away.

Disease Causing Fungi

Along with prokaryotes, fungi also form part of our microbiome, although they may not be as beneficial compared the prokaryotes. Most microscopic fungi living on our skin largely go unnoticed. In fact, it wasn't until we could easily sequence tiny fragments of DNA combined with the curiosity of a few scientists who went around taking swabs from peoples' skin that we learned we are covered in skin fungi. In fact, the average person may harbor between 30 and 60 different types of fungi at any one time. One of the most common types of fungal infections is athlete's foot, which causes dry flaky skin and irritation on feet. Other types of fungal infections cause dry and brittle toenails, which can be difficult to treat.

The rapid decline of amphibians in the last few decades has been one of the most tragic losses of an entire lineage of vertebrates in modern times. As we will learn in chapter 14, there are many reasons driving amphibian declines worldwide. But, one confirmed cause has been the spread the *Chytrid* fungus throughout the Central and South America. The *Chytrid* fungus has devastated entire amphibian communities throughout the tropics over the last several decades. Entire populations of frogs and toads have been wiped out within a few months after the arrival of the fungus.

In 2008, I traveled to a remote region of Panama with herpetologists from the University of New Mexico to find rare reptiles and amphibians. After my first couple of nights, I was happy to have found 10 species of frogs along a one-mile trail through the tropical rainforest. Over the course of a week of intense herping (that's looking for reptiles and amphibians) we managed to find 20 species total. Near the end of the trip, I learned from another colleague who had been there 1.5 years earlier that there were once almost 90 species of frogs and toads along the same trail. In a very short time, the *Chytrid* fungus spread to this area wiping out nearly 80% of the amphibian diversity in less than a year. Surveys in subsequent years have not found most of those missing amphibians.

Amphibians are highly susceptible to fungal infections on their skin because they use it to breathe and regulate electrolytes. The *Chytrid* fungus infects their skin rendering them unable to regulate their electrolytes, resulting in a quick death for the animal. Fortunately, the *Chytrid* fungus is not fatal to every species of amphibian.

Our food crops can be infected with a black colored fungus called ergot (*Claviceps*), which produces several toxic chemicals. The typical symptoms include nausea and vomiting and in some cases hallucination. Ergot has been documented infecting an entire town's wheat supply. In fact, the Salem Witch Trials, when the town "went mad", may have been caused by a large outbreak of ergot making people hallucinate.

Yeast and Civilization

Civilization has its roots in the agricultural revolution that allowed people to transition from a wandering hunter-gatherer life-style to taking up more permanent residence. Once people were free of having to find food everyday, they could pursue other interests. Within a short time, there were farmers, builders, poets, philosophers, scientists, teachers, politicians, and lawyers (that list is not meant to be in any particular order). Why did people abandon the hunter-gatherer lifestyle, what brought about the agricultural revolution?

Several hypotheses have been put forth to explain why our ancestors settled down and began farming. The fact that the agricultural revolution started about the time of the last ice age is not lost on anthropologists. Ever since the last ice age, humanity has greatly benefited from some of the most stable climates in the history of our species. As you might guess, the stable climate was a major factor in starting agriculture because you could predict when to plant and harvest your crops.

Another hypothesis to explain the start of agriculture centers on the evidence that our ancestors began planting and harvesting crops to brew alcoholic beverages including beer, wine, or mead. Ancient hieroglyphics from Egypt and other early civilizations clearly depict people drinking fermented beverages. The Egyptians used beer to pay workers building the pyramids a wage of one gallon of beer per day. Other ancient civilizations had it written in their laws that beer could serve as payment.

There were several benefits of drinking beer in ancient times. Diarrhea caused by various waterborne pathogens including *Giardia* and *Cholera* has been and continues to be a major cause of death in humans. Brewing beer or other alcoholic beverages would have reduced the risk of getting sick from drinking bad water.

Alcoholic beverages are made by using yeast to ferment some type of plant material. Yeast break-down the sugars found in plants; however, in the absence of oxygen, they make ethanol as a by-product of the fermentation process. Ethanol found in the beer kills many pathogens making it safer to drink than normal water. Additionally, beer provides a source of vitamins and carbohydrates for energy. Before you decide to give up on water and turn to a diet of beer, you should know that ancient beer was about 2-3% ethanol, compared to the much higher alcohol content upwards of 8% found in modern beers.

The other reason why beer may have been important for the rise of civilization lies in its ability to reduce social anxiety among people allowing them to more freely mingle and speak their thoughts. Basically, beer served as social lubrication and was an important part of feasts and celebrations that promoted social networking required in a civilization. But the main question that we may never know the answer to is, what prompted the first person to drink water with rotting wheat or fruit in it? I suppose it was some guys daring another guy to drink something gross.

Chapter 9

Most Animals Are Invertebrates

If aliens visited the Earth and randomly sampled its diversity, they would quickly learn that most animals are arthropods.

Introduction

If an alien race were to visit our planet and randomly sample the diversity of animals, they would quickly conclude that most animals are arthropods, an invertebrate. Traditionally animals have been divided into two broad groups; the vertebrates, animals with backbones, and the invertebrates, animals without backbones. From an evolutionary, point of view, this distinction is some somewhat arbitrary as starfish and sand dollars are more closely related to vertebrates than to other invertebrates.

Many of the animal phyla appear worm-like. Some, such as the peanut worms are rarely encountered, while nematodes (roundworms) are nearly ubiquitous and found in large numbers, but most are small and go relatively unnoticed. Some worms are parasites, and superficially they seem quite simple, yet have evolved extraordinary life histories as they move between different host completing complex life cycles. Sometimes these parasites even manipulate the behavior of their host, in some cases dooming them to die.

As you learn about animal diversity, keep in mind that all animals share a single common ancestor that lived about 600-700 million years ago. Therefore, all animals share traits, inherited from that last common ancestor. From there, animals have diversified into millions of species living today. Even more stunning is that modern diversity represents about 1% of all the species that has existed since the start of the Cambrian 542 million years ago.

By far, invertebrate diversity dwarfs the vertebrates, you can find them practically everywhere. Small crustaceans and snails live at the bottom of the ocean at depths exceeding 35,000 feet, while small spiders float miles high in the atmosphere using their webs to catch the wind. I start this chapter with a brief discussion of the evolution and diversification of animals. But, their incredible diversity makes it impossible to cover the entirety of animals in a single chapter. Therefore, I focus on just a few of the major groups of invertebrates to provide a glimpse of the stunning diversity of animals found in this world.

The Rise and Diversification of Animals

Life evolves and adapts as diversity gives rise to diversity. If the environment changes creating new opportunities, life finds a way to exploit them. But, for 3 billion years, life existed as single-celled organisms. Then, almost out of nowhere, the first animals appeared fully formed and rapidly diversified during an event called the Cambrian explosion. So profound was the rise of animals, that it ended the Proterozoic eon that spanned the previous 2 billion years and ushered in the Phanerozoic eon. This raises the question, why did it take so long for animals to evolve? The answers may be hard to find, but tough questions don't stop us from trying to find the answers.

Several problems make it difficult to know what events led to the origins of animals. Finding tiny fossils of cells in rocks billions of years old is hard. Also, the Earth is an active planet, continents move around, mountains are uplifted, and rocks with ancient fossils are eroded, removing any evidence they ever existed. The further we look back in time, the less evidence we have of early life and the environments they lived in. Despite these difficulties, theories explaining the sudden appearance of animals have been made.

Darwin, and other naturalists in the 1800s, noticed that animal fossils were abundant from Cambrian Period rock layers and above, but there were no animal fossils below these rocks. It's as if animals suddenly appeared in the fossil record fully formed. In Darwin's time, scientists lacked the tools or know-how to find microscopic fossils or to accurately age rocks. It took the discovery of radioactivity in the early 1900s and another 100 years of hard work to date rock layers around the world, placing a time line to the fossil record.

Rocks are dated using the steady radioactive decay of certain isotopes. Recall that all elements have isotopes, which are formed by varying numbers of neutrons in the nucleus. For example, carbon always has 6 protons, but it can have 6, 7, or 8 neutrons making carbon-12, carbon-13, and carbon-14, respectively. Some ratios of protons to neutrons are stable while others are not and decay into daughter elements with stable combinations of protons to neutrons. For example, carbon-14 (6 protons and 8 neutrons) decays into nitrogen-14, a stable element. Luckily for scientists, these radioactive isotopes decay into daughter elements at a predictable rate called the half-life. By using the ratio of radioisotopes to the daughter elements, an accurate age for a rock is determined. Most know about carbon-14, but every element contains radioactive isotopes. For ancient rocks, multiple elements including uranium and potassium are used to determine their age.

Once the age of the earliest animal fossils was determined, the next question focused on the causes driving the rapid rise of animals 542 million years ago. Below are the leading hypotheses to explain the appearance of animals at the start of the Cambrian. Keep in mind, that the actual answer may be a combination of these hypotheses.

Hypothesis 1: The Rise in Atmospheric Free Oxygen

Every rock holds clues to how and when it was formed. As we analyze the mineral composition and ratio of isotopes in rocks going further back in time, a picture emerges of an ancient Earth that lacked free oxygen (O_2) in the atmosphere for its first 2.5 billion years. Over time, through the constant action of photosynthetic organisms, oxygen levels slowly increased. By 2 billion years ago, oxygen levels were about 2-3% of the atmosphere, enough for bacteria to evolve aerobic respiration. Then about 600-550 million years ago, oxygen levels rose to about 12-15%. In a short time, geologically speaking, oxygen was six times more abundant than it had ever been.

Free oxygen in the atmosphere energizes life, it allows cells to extract 15 times more energy out of the same amount of food, thus enabling an active life style, a hallmark of animals. For this reason, the rise in atmospheric oxygen levels has been proposed as a major reason why animals evolved. Without oxygen in the atmosphere, the active life style of animals would not be possible. People who hike tall mountains have experienced sluggishness resulting from less oxygen at high elevations.

Hypothesis 2: The Evolution of Hox Genes

Animals are complicated organisms. As an animal, we are a colony of more than 10 trillion genetically identical cells working together for the common good of a single animal. Our cells may be genetically identical but, we have over two hundred types of specialized cells, including muscle, nervous, skin, *etc.*, to carry out the many functions of our bodies. Animals also have symmetry, meaning we have an organized body plan. Radial symmetry, or animals that look circular is the simplest body plan. More complex animals, like ourselves, have bilateral symmetry giving us a front and back, top and bottom, and left and right sides.

To develop from a single-celled zygote into a multicellular animal with a body plan, requires the coordination of numerous genes. Genes must be regulated, turned on at precisely the right time and then turned off again. To achieve this level of coordination in growth and development requires an entirely new type of master genes to coordinate growth and development at the genetic level.

The answer came in the form of *Hox* genes, which control the growth and development of animals from a single-celled zygote into an adult. They evolved early in the history of animals remaining relatively unchanged for millions of years. Although, some animals possess more *Hox* genes than others. Several interesting experiments show that these genes still function among different taxa separated by hundreds of millions of years. For example, if you insert human *Hox* genes into mice, or even the more distantly related fruit flies, these genes still work. It appears that development from a single celled zygote is a highly conserved process across all species of animals.

Mapping the location of *Hox* genes on chromosomes revealed they are organized in our genome based on their functions. Similar to fruit flies, we are segmented, a quick examination of our spine and ribs reveals a segmented body concealed by our skin. Each one of our vertebrae is numbered, indicating its position in our body. *Hox* genes line up with the segmentation in animals, coordinating growth and development by turning genes on and off at the correct time. Understanding the evolution of *Hox* genes and how they work is a major area of current research because it could provide breakthroughs in growing lost limbs, or damaged organs.

Hypothesis 3: The Rise of the Predators

A favorite hypothesis among paleontologists to explain the sudden appearance of animals is the rise of the predators setting off an evolutionary arms race. The closest living relative to animals is a single-celled eukaryote called a *Choanoflagellate*. How choanoflagellates evolved into the first animals probably went something like this. These tiny cells faced the dim prospect of being eaten by other eukaryotes. To avoid predation, several *Choanoflagellates* joined together forming a small colony too big to be eaten. By forming a small colony, individual cells could specialize in certain tasks allowing the colony to pursue new resources, including eating other eukaryotes or other colonies of eukaryotes. And so, an evolutionary arms race began, quickly driving the evolution of animals and their rapid diversification.

From this starting point with small colonies of cells, you could imagine the evolutionary arms racing continuing. Prey evolve to be larger, thus avoiding being eaten, predators evolve to be larger to go after the larger prey. The prey evolves legs or fins allowing them to escape predators, the predators evolve better fins and stronger muscles allowing them to swim or walk faster, or to become more agile. The prey evolves a hard outer shell to protect themselves from being eaten, the predator evolves teeth, jaws, or claws to break the shell. The predator evolves eyes to detect the prey, the prey evolves eyes to detect and avoid predators, and so it goes. In fact, eyes are so useful, it is estimated they evolved more than 20 times in animals.

I suspect a combination of all three of these events occurring at about the same time creating a perfect storm for evolution, triggering the rise of animals. With more free oxygen available, animals became larger and more mobile fueling the evolutionary arms race. The presence of *Hox* genes allowed for multicellular life to quickly diversify because it only takes small changes in *Hox* genes easily modify body plans and limbs. As a result, animal life quickly diversified into many different forms. By the end of the Cambrian Period 499 million years ago, most modern animal phyla, including arthropods, echinoderms, mollusks, cnidarian, and chordates, had all arrived and were diversifying.

What Makes an Animal Unique?

All modern species descended from ancestral species by accumulating modifications over time adapting to their environment. By studying different animals and comparing their similarities, we can piece together the characters shared by the last common ancestor of all animals. Taken together, the similar features of animals define the kingdom Animalia as a unique group, different from all other organisms. Below is a description of the common features of animals.

In an introductory and high school biology course, you may have suffered through the rite of passage of dissecting some fetal pig or feral cat. But, getting past some squeamishness of cutting open an animal to reveal its insides, the organization of the organs and anatomy reveals itself. It turns out that our organs and tissues arise from three different germ layers during development. In fact, all animals with bilateral symmetry have three germ layers, the endoderm, mesoderm, and ectoderm. The endoderm forms our digestive tract including the esophagus, lungs, stomach, liver, and the colon. Our mesoderm forms muscles, bone, connective tissue, fat tissue, and circulatory system including the heart. The ectoderm forms our skin, teeth, hair, and parts of the nervous system.

Symmetry in animals comes in two types: radial symmetry and bilateral symmetry. One type of animal with radial symmetry and two germ layers (endoderm and ectoderm) are the Cnidarians. Despite their simple body plan, cnidarians form diverse and important group including sea anemones, jellyfish, and corals. All other animals have bilateral symmetry; front (dorsal), back (ventral), left and right, top (anterior) and bottom (posterior) ends. Bilateral animals move through the environment with their head first. As a result, most of their sense organs including, eyes, nose, ears, and antennae are on the head.

Most animals move through their environment, whether using legs to walk, wings to fly, or fins to swim. Therefore, all animals have muscles allowing them to move. Some animals such as barnacles gave up their ability to move as adults, but they still use muscles in their legs to collect plankton passing by. To coordinate muscle movement, animals use a nervous system, another unique feature of animals allowing for rapid long-distant communication. Just think how quickly you can react to touching a hot stove!

Lastly, animals are heterotrophs that feed by ingestion. Animals are the most active organisms on the planet. To power their active life style, animals require lots of energy and eating other organisms provides the necessary energy to fuel animals. The high energy demands of animals are one reason why they can't be totally photosynthetic like plants. However, if there is one rule in biology, it's that some animal, somewhere, has evolved some way of defying the general trends. As we will learn, some animals have lost their mouths while other animals are photosynthetic!

When it comes to sponges, I don't consider them animals. Although, almost every classification system includes sponges as animals. There are several reasons for this; sponges are multicellular organisms with specialized cells, they feed by ingestion, and molecular data, and specialized collar cells cell closely resembling choanoflagellates clearly suggests sponges shared a common ancestor with all animals. In some ways, they represent the transition from cellular colonies with specialized cells to animals with specialize cells, true tissues, and symmetry. However, because sponges lack true tissue layers, symmetry, muscle and nervous tissue, I consider them to be "para-animals" or sister to animals. But, not everyone agrees with me.

Cnidarians
Stinging Tentacles and Reef Builders

Cnidarians are considered to be "primitive animals" due to their simplistic body plans and lack of a brain. They have radial symmetry, appearing circular. If you were to draw a line through the middle, the two halves would be mirror images of each other. With two germ layers, an ectoderm and an endoderm, and lacking a head region, they resemble the most ancient animal fossils found. Their name comes from the term cnidocytes for their stinging tentacles, a character shared by all cnidarians.

I'm always a little hesitant to call animals primitive, Cnidarians have just as long as an evolutionary history as us humans. Despite remaining simple, their body plan has worked well for 600 million years, changing little in all that time. They are a great example of natural selection acting to maintain a successful body plan. Today, Cnidarians remain a diverse animal phyla with about 10,000 species in four major classes: the jellyfish (Scyphozoans), corals and anemones (Anthozoans), sea wasp and box jellies (Cubozoans), and the Hydrozoans including the tiny hydra and the much larger Portuguese Man-O-War.

Most of us are familiar with jellyfish. They belong in the class *Scyphozoa*, which means 'cup-like animal'. These free-floating animals swim by contracting their bell. Many species use light sensitive pigments to guide them through daily vertical migrations in the water, ascending from the depths towards the light of day. So far, about 200 species of jellyfish have been discovered throughout the world's oceans from the tropics to the poles.

The largest jellyfish is the Lion's mane Jellyfish (*Cyanea capillata*) found in cold arctic waters, it can reach 7.5 feet in diameter with its tentacles trailing nearly 40 feet behind it. Adult jellyfish of smaller species are a favorite of the leatherback sea turtle who sometimes eat plastic bags, mistaking them for a jellyfish. In recent years, overfishing has drastically reduced fish populations, which are natural predators of jellyfish. As a result, some populations of jellyfish are exploding, especially in the Sea of Japan. To make matters worse, the large numbers of jellyfish are eating the babies of the same fish being harvested, exacerbating the rapid decline in fish populations.

While many jellyfish can inflect annoying stings, most aren't deadly to humans unless you're allergic to them. This is not the case for the sea wasp or box jellies (class: Cubozoan). Although, superficially resembling jellyfish, they are not closely related to them. They get their name from their box-like appearance. It's difficult to rank the deadliest animals in the world; there are many capable of quickly killing someone, but several cubozoans would certainly be near the top of anyone's list. The venom from a single sea wasp (*Chironex fleckeri*) could kill up to 60 adults within 4 minutes! Luckily for us, these are not aggressive animals. Their toxin has evolved to quickly kill their prey, mostly small fish, thus preventing them from being damaged. Fortunately, fatal encounters with sea wasps are rare in the Indo-Pacific. While most Cnidarians detect light with light-sensitive cells, sea wasps use sophisticated eyes capable of forming an image, so they can hunt down their prey.

Another common Cnidarian are the sea anemones in the class Anthozoa, which literally means 'flower animals'. They are well known to the public, especially since the movie *Finding Nemo* correctly portrayed clownfish (*Amphiprion*) in a symbiotic relationship with sea anemones. About 1,000 species of anemones inhabit the oceans, and like all Cnidarians, sea anemones use stinging tentacles to capture their prey. Clownfish evolved close relationships with about 10 species of sea anemones in the Indo-Pacific regions. In this relationship, both species benefit; the anemone protects the clownfish from predators and the clownfish cleans the anemone of waste and potential parasites. To prevent them from being stung, clownfish produce a thick layer of mucus.

A close relative to sea anemones, corals also deserve the name "flower animals". Similar to flowering plants, they come in a variety of shapes and brilliant colors rivaling any terrestrial ecosystem. Pigments made by the coral to protect them from UV light make many species brightly colored. Worldwide, more than 5100 species of corals live mostly in shallow waters. Although, the true extent of their diversity has yet to be realized because different species look similar. Cryptic species closely resemble each other in appearance due to convergent evolution, but may not be closely related. Recent molecular work suggests there could be many more species yet to be discovered.

Corals are among the most important organisms in the oceans for their role in creating diverse ecosystems. These small animals form large colonies by making rock from dissolved minerals. Overtime, the rocks grow forming large coral reefs in the tropical regions of the world. While coral reefs account for less than 1% of the ocean's area, it has been estimated that nearly half the ocean's diversity may be found or depend on coral reefs. Similar to trees providing additional habitat on land, coral reefs provide many habitats in regions that would otherwise be a biological desert.

There's a paradox of coral reefs; they are among the most diverse and productive ecosystems in the world, yet they exist only in crystal clear water. The paradox lies in the fact that clear water lacks nutrients required by organisms to grow, including phytoplankton. Phytoplankton, being autotrophic, forms the base of most marine ecosystems supplying nutrients and energy to all the other animals. However, most species of corals require lots of sunlight to grow. Nutrient rich waters promote algae and phytoplankton growth reducing sunlight and promoting algae growth on the corals, both these factors damage corals, or even kill them. Yet, coral reefs thrive in nutrient poor conditions, rivaling tropical rain forests in diversity and ecological complexity.

The success of corals in nutrient poor waters depends on their symbiotic relationship with tiny photosynthetic eukaryotes called zooxanthellae. These tiny organisms live inside the coral's cells at densities approaching a million cells per square inch. Here's another example of a symbiotic relationship in nature when both organisms benefit from the arrangement of two organisms living together. The corals send out their small tentacles collecting tiny crustaceans and plankton floating in the water, and the zooxanthellae produce carbohydrates to supply the corals. If there are too many nutrients in the water, then algae may overtake the corals killing them. In recent years, coral reefs have suffered major losses due to a variety of problems including global climate change, pollution, and over harvesting.

Mollusks
From the Primitive to the Advanced

What does a scallop, octopus, and a snail have in common? They are mollusks, an incredibly diverse group of animals with over 100,000 species. Most mollusks have a shell made of calcium carbonate, similar to the reef-building corals. However, several groups of mollusks have

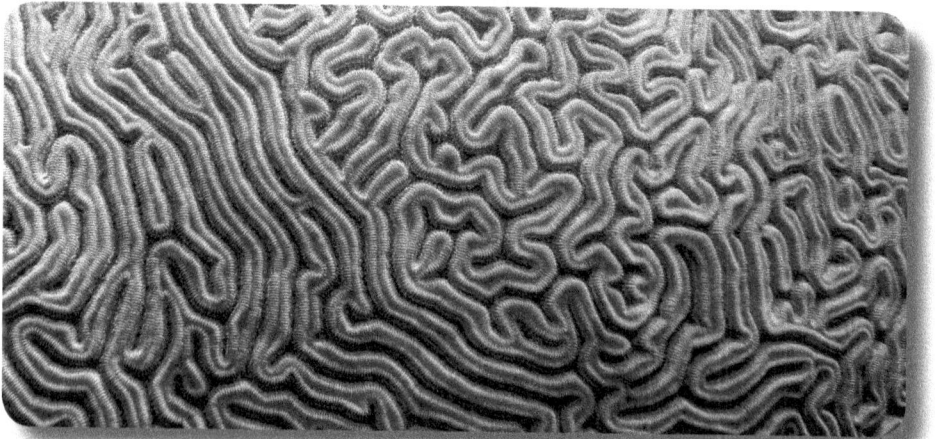

A closeup image of a brain coral, a colony of tiny animals

independently lost their shells, including squid, octopus, garden slugs, and sea slugs.

Even with the incredible diversity of mollusks, they share several traits in common, inherited from their last common ancestor over 500 million years ago. The most obvious feature is a large body cavity called the mantle containing their organs and some use it for breathing. Except for bivalves, mollusks have a radula, a unique mouth part modified in each class to accommodate their feeding habits. In most snails, the radula scrapes algae from rocks and other hard surfaces. In cephalopods it is modified into a horny beak made of keratin and used to eat fish and crabs.

Most people are familiar with snails, animals that make a protective shell made of calcium carbonate. In marine environments, snail shells can be large, colorful, and covered in unique patterns. An interesting aspect of snail shells is that their shape and color patterns follow precise mathematical sequences such as the Golden Mean or Fibonacci sequences. Snails have varied lifestyles, many eat detritus, aquatic plants, and algae, while others are predaceous. Most snails are harmless, except for the cone snail (*Conus*), which is one of the deadliest animals on the planet. A tiny amount of the neurotoxic venom can quickly kill an adult. Scientists are currently studying their venom to learn how our own nerve cells work.

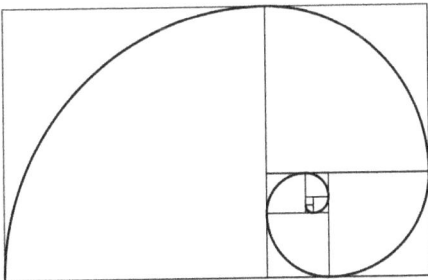

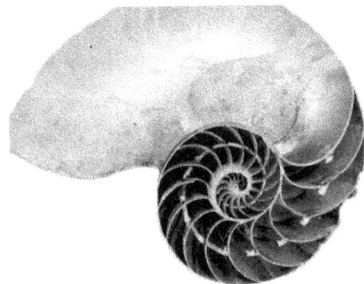

The chambered nautilus uses the Golden Spiral for its shape

Some snails gave up their shells like the common garden slug and the highly colorful marine nudibranchs (nudi = naked, branch = gills or naked gills), also known as sea slugs. The bright colors of nudibranchs serve as a warning to other animals not to eat them. Many are poisonous or taste bad, so predators leave them alone (same as in monarch butterflies and coral snakes).

A relative of the sea slug, *Elysia chlorotica*, eats algae at a very early age and then does something quite remarkable; it doesn't fully digest the algae. Instead, it incorporates the chloroplasts into its own cells, looks green, and is capable of photosynthesis. It can go a long time without feeding when sunlight is abundant. Additionally, its cells support the chloroplasts keeping them photosynthesizing for months.

Bivalves, or animals with two shells include commonly seen animals that often end up on our dinner plate including mussels, scallops, and oysters. Similar to snails, their shells are made of calcium carbonate, but these animals lack a head and the radula. These animals are typically sedentary or even sessile filter feeders. Because of their life style, most need only a simple nervous system and lack eyes. However, bay scallops can swim to avoid predators and use many small eyes to see predators.

The cephalopods are an interesting class of mollusks including the chambered nautilus, squid, cuttlefish, and the octopus. By several standards, cephalopods are highly advanced. With the exception of the chambered nautilus, most have lost their shells. By losing their shells, they lost protection, but they gained speed and maneuverability. Cephalopods include the largest of all invertebrates, the colossal squid (*Mesonychoteuthis hamiltoni*), found in the southern oceans. It reaches a length up to 46 feet and weigh up to 1600 pounds. Very little is known about its life history and other habits. It is believed to be an ambush predator of deep-sea fish, in turn, it is prey for sperm whales.

The octopus has been at the center of a controversy over its perceived intelligence. Judging and measuring intelligence in other animals is a difficult task confounded by the fact that there is no universal test of intelligence and many researchers disagree over what factors are important for determining intelligence. In the wild, the octopus is an active predator, often eating crabs. When hunting, some octopus will locate a crab and wait to ambush it when it emerges from its hiding place, indicating forethought. Interestingly, some cephalopods actively change their skin color for communicating emotions, confusing their prey, and of course they change color for camouflage. Squid are also known to participate in cooperative hunting for fish.

To top it off, cephalopods have achieved these abilities with a very different nervous system and brain structure compared to vertebrates. Additionally, the cephalopod eye evolved independently of the vertebrate eye, but works the same way and with a similar level of acuity.

The Arthropods: A Body Plan for Success

It's difficult to determine which group of animals is the most successful, it may simply depend on your criteria. If you were to choose success based on diversity and sheer numbers of individuals, then arthropods would claim the title as the most successful group of animals on the planet. To a first approximation, the average animal is an arthropod. Currently, there are more than a million described species of arthropods with new species constantly being discovered. Perhaps, their modular body plan of a hard exoskeleton and jointed limbs accounts for their success. It doesn't take much change in an arthropod's DNA, only a few small mutations, can lead to changes in their body, quickly leading to a new species. When combined with rapid generation times, arthropods quickly evolving to adapt to a changing environment or to exploit a new resource.

The arthropod exoskeleton is made of a tough substance called chitin, the same chemical that forms the cell wall of fungi. All arthropods routinely shed their exoskeleton, allowing them to grow each time they molt. Despite the success of arthropods, the exoskeleton, places an upper size limit on them. Luckily, we will never have to face a giant 50-foot praying mantis, the result of mutations from radioactive fallout as portrayed in 1950s science fiction films. As an animal gets larger, so must its support, but the scaling is not linear. Meaning, if I were to take a mouse and enlarge it proportionally to the size of an elephant, its bones would instantly break from its own weight. If arthropods were to grow much larger, their exoskeletons would have to rapidly increase in thickness. At some point, the exoskeleton would be so large, that the animal could not move. Despite the size limitations on arthropods, some species have become relatively large, especially in prehistoric times.

The first arthropods colonized the land about 425 million years ago, and by 400 million years ago, giant centipedes 6 feet long roamed the land. In the ocean, giant sea scorpions reaching 7 feet long swam in warm coastal waters preying on fish and marine invertebrates. These ancient arthropods were able grow larger, potentially reaching the upward limits of their size, because there was more oxygen in the atmosphere. Arthropods obtain oxygen through simple diffusion; if the animal grows too large, then it won't get enough oxygen to all its internal organs and muscles. But, 400 million years, there was about 8% more oxygen in the atmosphere, enough to let those ancient arthropods grow to their maximum possible size

Arthropods are well known for their compound eyes made up of tens to thousands of individual lenses called an ommatidia. Their eyes work much like pixels in a digital camera, resolution improves with more pixels. In arthropods, the more ommatidia means the better their vision. Contrary to what's portrayed in the movies, arthropods do not see multiple images, it's more like a low-resolution photograph.

There are four major groups of living arthropods, the myriopoda (centipedes and millipedes), the chelicerates (horseshoe crabs, and arachnids), the crustaceans (shrimp, crabs, and many others), and the insects, the most species-rich group on the planet. Trilobites represent a fifth group of arthropods that existed for nearly 300 million years before their extinction 251 million years ago.

Chelicerates
Horseshoe Crabs and Arachnids

The chelicerates include many familiar arthropods including spiders, ticks, scorpions, and the prehistoric looking horseshoe crabs. They get their name based on the chelicerae, an appendage found in front of the mouth and shared by all members of this group. As you may have guessed, they inherited this feature from a distant common ancestor that lived around 500 million years ago.

In most chelicerates the chelicerae are simple pincers used for feeding. In spiders they are modified into fangs used to inject venom into their prey. All Chelicerates, including Arachnids lack antennae, easily distinguishing them from insects and crustaceans. They have two main body parts, the abdomen and a cephalothorax, which is a fusion of the head (cephalo) with the thorax (head – abdomen). There are three major classes of chelicerates; the horseshoe crabs, sea spiders, and arachnids, which are the most diverse lineage including spiders, scorpions, wind scorpions, and ticks.

With fossils dating back 450 million years, the horseshoe crabs (*Limulus polyphemus*) are among the oldest living species. They are not related to true crabs, but are more closely related to spiders based on the presence of chelicerae and the cephalothorax. Fossils of ancient horseshoe crabs show that they have changed very little in 450 million years, they really are a living fossil.

You can find horseshoe crabs on the Atlantic and Gulf coasts where they emerge from the surf during by the thousands during spring high tides to lay their eggs. Each year, as they have done for hundreds of millions of years, horseshoe crabs will lay their eggs all at once right at the top of the high-water line. Shorebirds time their migrations to coincide with the egg laying to take advantage of the nutritious eggs. In recent years, horseshoe crab populations have dwindled solely due to over-harvesting. The loss of horseshoe crabs and their eggs has resulted in significant losses of shorebird populations, especially the Red Knot.

Perhaps the most feared animals on the planet, most arachnids live on land. It's been estimated that 52% of all women and 35% of all men suffer arachnophobia, the fear of spiders. This is unfortunate because very few spiders can harm humans, and spider bites from those species are quite rare. Importantly spiders regulate insect populations making them a vital part of most terrestrial ecosystems because they are voracious insect predators. Although, some spiders are large enough to prey on small rodents, birds, and the fishing spider lives at the edge of water and preys on small fish.

As you already know, many species of spiders build intricate webs to catch their prey. They use silk made of proteins, which is stronger than steel, to spin their webs. Because the webs are sticky and collect dust, reducing their ability to trap insects, the spiders will eat their own web to recycle the proteins, only to continually rebuild them. Not all spiders build webs to catch prey, small jumping spiders in the family Salticidae, which literally translates to "jumping", are active predators constantly on the hunt for prey. They have two large forward facing ocelli, which function as eyes giving them great depth perception, similar to humans. In the late fall, when outside temperatures begin to drop, many spiders will move indoors. Many of these spiders, including the small jumping spiders are actually beneficial because they don't carry diseases, don't build webs, and will catch numerous unwanted small insects in your kitchen and home, including roaches.

Other arachnids include the scorpions and wind scorpions. Scorpions are uniquely shaped predators with a long and narrow abdomen modified with a stinger at its tip. They can deliver a powerful sting to paralyze its prey. Wind scorpions are among the fastest of all arachnids, using their speed to chase down their prey. Some species in Middle Eastern deserts, known as camel spiders, can reach 6-inches. Wind scorpions have the unique distinction of possessing the largest jaws compared to its body of any animal. While not venomous, their jaws are over twice the size of their head allowing them to inflict a powerful bite as they shred apart their prey. Ticks and mites are ectoparasites, feeding off the blood of other animals. Unfortunately, some ticks carry harmful diseases.

Crustaceans
Include the Smallest and the Largest Arthropods

Copepods, less than a mm in length to American lobsters weighing 50 lbs, crustaceans include the largest and smallest of all living arthropods. The vast diversity of crustaceans is found in the oceans, but a few are found in freshwater lakes and rivers, and the familiar pill bugs are commonly found on land throughout much of North America. Crustaceans are somewhat similar to insects because both groups have antennae. Crustaceans have more legs than insects along with appendages on their abdomen, and crustaceans never developed flight. Despite that, crustaceans are an incredibly diverse lineage with over 67,000 known species. They include the tiny copepods, sessile barnacles, primitive looking fairy shrimp, and the more familiar crabs, shrimp, and lobsters.

Tiny copepods are among the smallest of all animals, and about 13,000 species are found almost anywhere including the oceans, lakes, and rivers. They are incredibly abundant and may account for up to 80% of the zooplankton mass in some areas. Estimates of their population predict there could be 1×10^{15} copepods in the world, or a thousand trillion individuals, making them the most abundant animal on the planet. In one the most profound experiences of my life, I went 2200 feet deep in a submarine, the Johnson Sea Link, to a deep-sea brine pool. For the entire 45-minute trip to the surface, I watched a steady stream of copepods float by the window. I too came to the same conclusion that copepods were incredibly abundant, to see so many animals in a random spot in the Gulf of Mexico, only to realize they would be relatively evenly distributed throughout the world's oceans.

Copepods are not the only small crustacean, in the desert southwest during the summer monsoon season, small temporary ponds fill with water and quickly become populated with tadpole shrimp (*Triops*), another living fossil whose form has not changed in 300 million years. In a matter of weeks, they will hatch, grow, mate, and lay their eggs, quickly completing their entire life cycle before the ponds dry. When the ponds dry up, strong winds blow their eggs for many miles spreading the species over large geographical areas. These ephemeral ponds are also home to small fairy shrimp, a diverse lineage of small crustaceans that appear to swim upside down. Tadpole shrimp and fairy shrimp resemble some of the earliest crustaceans dating back to the Cambrian. Perhaps, the reason they still survive today is by living in ponds without fish that would quickly eat them.

Perhaps the best-known crustaceans belong to the Decapods (deca=ten, pod=foot), which includes shrimp, crabs, and lobster. Based on this list, it becomes apparent why we know them, for some people, they are quite tasty. The largest living arthropod is the giant Japanese spider crab (*Macrocheira kaempferi*) with a leg span measuring almost 18 feet across. Living in mostly marine environments, crabs come in a variety of shapes and sizes. Hermit crabs carry snail shells on their backs for protection. Sometimes, small sea anemones hitch a ride on the shells to camouflage the crabs from predators and provide them with food. Small hermit crabs often go about trading each other's shells several times a day. At night, or during the day, small white crabs scurry on sandy beaches around the world. Known as ghost crabs, they quickly dart into their holes if approached too quickly.

Shrimp also come in many sizes shapes and colors. Small krill, less than an inch in length serve as the main food for whales, the largest animals on the planet. Inhabiting tropical reefs, is a striped shrimp with two large pairs of white antennae, these are the cleaner shrimp (*Lysmata amboinensis*). At cleaning stations, fish will line up allowing the cleaner shrimp to enter their mouth and gills removing any visible parasites from the fish. The relationship is mutually beneficial for both the shrimp and the fish, the fish gets pesky parasites removed and the shrimp gets a meal. Another small shrimp called a pistol shrimp has a unique claw that makes a loud snapping sound. They form a mutualistic relationship with small fish called shrimp gobies. The pistol shrimp digs a hole where they both live and the goby, with its good eyesight, protects the pistol shrimp.

Considered a delicacy by many, lobsters are also a decapod. In North America, there are several species; the American lobsters (*Homarus americanus*) with two large claws and are one of the more commonly encountered commercial species. In Florida and the Caribbean, Spiny lobsters (*Panulis argus*) lack large claws and are known for large migrations to their spawning grounds. They will follow each other in single file lines, maintaining contact through their antennas. Lobsters live up to 70 years, but unlike us, they don't die of old age. Their life span is more a function of growing too large, the larger the lobster, the more energy it takes to molt. Eventually, they grow so large that the energy expenditure to molt kills them through exhaustion. Another interesting fact about lobsters is they don't really grow old and senesce similar to humans. We don't know exactly why, but one reason could be that they express telomerase as adults, which means that there is no limit to the number of times their cells can reproduce.

Related to decapods are about 400 species of the colorful mantis shrimps. These voracious predators have two unique adaptations. First, their large eyes are among the most complex visual systems in the animal kingdom with 16 different photoreceptor pigments, compared to three photoreceptors in our eyes. Second, they have incredibly powerful raptor-like claws that move so fast it sends a shock wave through the water with enough power to stuns or kill their prey. Mantis shrimp attack animals with hard shells and some reports that they can break the glass of home small aquariums.

Nature has an Inordinate
Fondness for Insects

Insects are found on every continent and are by far the most diverse animals with over one million species currently described. They originally evolved on land over 400 million years ago during the Devonian, before vertebrates colonized land. Insects have continually been diversifying ever since. Flight evolved one time in all the invertebrates, and it was the insects that did it. With their diversity has come many novel adaptations in the animal world. Excluding humans, ants, bees, and wasps form the most complex societies on the planet. Butterflies and beetles undergo complex metamorphosis where larvae and the adults look and behave differently from each other. Other insects have aquatic larvae and the adults spend their life on land.

For the remaining part of this chapter, I cover some of the fascinating features of a few insects, just enough to provide a tiny glimpse into their diversity. At the end of the chapter, I will reveal the most dangerous insect in the world.

What's the best predator in the world? It could be the dragonfly, based on an incredibly efficient catch rate of 90-95%, far better than other well-known predators such as cheetahs (about 50% success rate). Another interesting feature of dragonflies is that they have changed little since their first appearance in the fossil record about 325 million years ago, it seems as if nature got their design right from the start. Today, there are about 3,000 species of dragonflies.

They have large eyes for an insect, allowing them to see small prey on the wing. Like other insects, they have six legs, but they aren't used for walking. Instead, they capture flying insects. With two pairs of wings capable of independent movement, make dragonflies some of the best flying animals in the world. In fact, dragonflies are so good at flying, the U.S. military is currently studying their wings to develop new flying technology. Larval dragonflies are adept hunters as well. Unlike the adults, larval dragonflies live in aquatic environments, ambushing their prey, including small fish using a modified mouth called the prementum that shoots forward.

Stoneflies, mayflies, and caddisflies are similar to dragonflies with winged adults and larvae confined to aquatic habitats. Mayflies belong to the order Ephemeroptera, which means fleeting (ephemera) wing (ptera). The larval stage of mayflies may last for a couple months to several years. Then, as if on cue, all the larvae will crawl out of the water, and molt into adults by the tens of thousands at the same time. The adults do not feed, and must find a mate before they die, all within a day. By timing their emergence all at once, they can overwhelm their predators and increase their chances of mating.

Caddisflies larvae make casings out of small pebbles, leaves, or debris from their environment. They are related to the butterflies and moths, which is evident by the silk they make to form their casings, and the adults are easily mistaken for small moths. Stonefly larvae are commonly found in clean running waters with caddisflies and mayflies where they are a favorite food of trout. Relying on clean, pollution free waters, these three insects make good indicators of stream quality because most of them are adversely affected when water quality declines.

There is a group of insects literally called the true bugs. With somewhere between 50,000 and 80,000 species including such different insects as aphids, cicadas, and assassin bugs. Despite this diversity, they all possess a modified mouth called a rostrum used for piercing and sucking. Aphids, small bugs found on many garden plants, insert their rostrum into the plant's phloem. The phloem is part of a plants vascular system used to transport sugars from the leaves to the rest to the roots. Aphids can be seen lined up on plants with their heads down, constantly drinking from the plants. If you look closely, you will see ants tending and protecting the aphids. The aphids supply the ants with sugar water, in turn, the ants protect the aphids, another example of symbiosis. Many true bugs are predators, including assassin bugs which hunt insects on flowers and use their rostrum to suck fluids out of their prey.

During summer nights, the forests come alive with sound of cicadas. They spend most of their lives as nymphs underground, one species even spends 17 years in this stage. Eventually, the nymphs will leave the safety of living underground, crawl up a tree, and molt into adults. Their song is produced from modifications in their abdomen, which is mostly empty space.

Butterflies and moths form one of the most recognizable groups of insects in the word. In North America alone, there are over 800 species of butterflies and nearly 10,000 known species of moths. These insects undergo a process called complete metamorphosis when the larvae restructure its body as they transform into an adult. Butterflies and moths both form a cocoon to protect themselves during this vulnerable time. The reason for this radical change in body form is an evolutionary strategy to prevent the adults from competing with the larvae over the same resources. For example, butterfly caterpillars feed on leaves, whereas the adults feed mostly on nectar.

Many species of moths and butterflies lay their eggs only on certain species of plants that they have been co-evolving with for millions of years. Unable to move rapidly, caterpillars need defenses to protect themselves from predators and parasitoids; including camouflage, bitter taste, poisonous hairs and various projections that inflect an annoying sting or irritation. Butterflies and moths are also a prime example of adaptive radiation, which is an increase in diversity that coincided with the evolution of flowering plants.

Often feared by humans for their powerful stings, ants, bees, and wasps are fascinating insects that includes some of the most complex societies on the planet and unique mating systems. Ants are one of the most important insects in the world and serve many important functions. For example, their burrows aerate the soil and other species disperse seeds. Some species form large colonies, sometimes exceeding a million or more individuals. About 200 species of army ants found in the tropics form mobile colonies carrying all their eggs and the queen as they move through the jungle. They quickly invade an area, known as a raid, where they will forage and eat any small animal in their path that cannot get away from them.

With their division of labor and large size, leaf cutter ants form some of the most complex societies outside of humans. They evolved the ability to farm fungi long before humans. Large workers can be seen constantly bringing leaf parts into the colony, which are used to cultivate and grow a particular fungus eaten by the ants. Small workers, ride on top of the leaves carried by the much larger workers, to protect the larger workers from tiny parasitoid flies and wasps. Other workers remove waste from the colony while others tend the fungus gardens.

Bees and wasp are feared for their powerful stings from a modified body part called an ovipositor. Not all wasps form large colonies, some are solitary and use their ovipositor to lay their eggs in other animals or plants. These parasitoids are slightly different than a parasite. Parasites typically don't kill their host. But, a parasitoid will. Tarantula Hawks, one of the most well-known parasitoids lives in the desert southwest. These large solitary wasps hunt tarantulas to lay their eggs. Upon finding a tarantula, a tarantula hawk will sting the spider, not to kill it, but to paralyze it. Once it's paralyzed by her stings, she will then drag the spider down its own hole, lay her eggs in it, and then bury the spider, alive. The eggs hatch and the larval wasp eats the spider from the inside out, only to kill it once the wasp emerges as an adult.

When it comes to diversity, beetles reign as the king, they are the most diverse order of insects on the planet. Some estimates indicate that one out of every four animal species is a beetle! Similar to butterflies and moths, their rise in diversity in the last 100 million years coincided with the evolution of flowering plants. Beetles live everywhere, and many species have yet to be described by scientists. In a famous experiment to determine insect diversity, one scientist fumigated a single tree in the tropics to estimate how many insects were in the canopy. Upon collecting all the insects, they discovered about 50 new species of beetles from this one experiment making you wonder how many more undescribed species remain in that single forest. Like all other groups of diverse animals, beetles can be both beneficial and harmful to humans, some species are crop pest, while others like the lady bugs are predators of crop pests. Other beetles are important pollinators.

Flies are another very diverse order of insects with most species being completely harmless and many are beneficial. Unlike most other insects, flies have a single pair of wings and the second pair is reduced to small halteres that function much like a gyroscope. Fortunately, not all flies are human pests. The small fruit fly has been used for over a hundred years as a model organism to study genetics. Flower flies are harmless and important pollinators in many ecosystems, some mimic local bumblebees so that predators will leave them alone. However, robber flies are ambush predators of insects, duping their prey by mimicking harmless bee.

What is the most dangerous insect on the planet? If you ask someone who studies diseases, they will tell you it is mosquitoes, a type of fly. It's not that mosquitoes are super venomous or produce a super toxic poison. Instead, their danger lies in the fact that mosquitoes and other flies transmit human diseases, including malaria, west Nile virus, dengue fever, and yellow fever. It has been estimated that each year 700 million people get some disease from mosquitoes, killing over one million people.

Echinoderms
Relatives to Chordates

It may be hard to believe, but starfish and sea urchins, seemingly simple animals in the phylum Echinodermata, are the closest living relatives to chordates, the phylum which includes us. The close relationship was originally based on similarities in early growth and development, and later further supported by DNA evidence. Echinoderms are a varied phyla with familiar animals including, sea urchins, sea cucumbers, starfish, and sand dollars. Their radial symmetry is only superficial, in fact the ancestors to star fish were bilateral, and so are their larvae. I often wonder what caused these animals to forsake bilateral symmetry and evolve radial symmetry.

About 7,000 species of Echinoderms live in marine environments from shallow coasts to the depths of the oceans and often in great numbers. Similar to cnidarians, echinoderms lack eyes, a head, and centralized sense organs. However, the tube-feet of starfish and sea urchins contain a light-sensitive protein called opsin, allowing them to detect areas of dark and light. We have opsin in our eyes as well, which is used for the same purpose. Sometimes, you may see a starfish with one of their arms slightly lifted in the direction they are moving.

Some starfish live up to thirty-five years. Although they reproduce sexually, they can also reproduce asexually by physically detaching one of their five arms. The starfish will then regrow the arm, but even more remarkably, the severed arm will also grow into a new starfish! Another fun fact about a starfish is that they eat with their stomach inside out. When a starfish approaches a bivalve, it uses its arms to pry the shells open, then forces its stomach into the prey to digest it. This way the starfish can eat larger prey without having to actually ingest the whole animal.

Sea urchins graze on marine algae, often reaching very large numbers. In the pacific coast of California, sea otters keep sea urchin populations in check. Sea otters live among kelp forest where they dive down in search of urchins. Once they find one, the otter will return to the surface, float on its back to crack it open and eat the tasty insides. By eating the sea urchins, otters make it possible for the large kelp forests to exists, thus greatly increasing diversity of the near-shore marine ecosystems along the west coast of North America, from California to Alaska.

Sea cucumbers are less well known than sea urchins and starfish, but they possess a unique talent. When threatened with predation, they will eviscerate themselves and then regrow their insides!

The Worm-like Animals

There are many other major groups of animals, most appear worm-like, lacking appendages, and include various parasites that continually plague mankind. Parasitic worms may look simple, but many have complex life cycles where they infect several hosts like snails and birds, or snails and humans. To get from one host to another, these parasites will alter the behavior of their host, often killing it in order to infect the next host. Unfortunately, it is impossible to cover their diversity here. But, I have included a brief description of the larger groups.

The flatworms worms get their name because they are flat, an ideal body shape well suited for a parasitic lifestyle. They include liver flukes, blood flukes, and tapeworms. By maximizing their surface area, flatworms lack a mouth and absorb all their nutrients across their skin. Tapeworms, a common parasite, live inside the gut of vertebrates. If you eat raw or undercooked meat, you could become infected with a tapeworm, just think over the next 30 years it could grow up to 50 feet long, producing millions of eggs released whenever you go to the bathroom.

Some parasites change the behavior of their host, in some cases causing it to commit suicide to perpetuate the life cycle of the parasite. Horsehair worms are parasites of insects resembling the coarse hair of a horse's mane. Any insect infected by a horsehair worm is doomed to die. Unsuspecting insects will eat the eggs of the worm near water. Once ingested, the eggs hatch, then grow inside the abdomen until the worm reaches maturity. Then, it will cause the insect to jump into water, drown itself, then the worm will burst out of the insect's abdomen into the water. Once in the water, they will find a mate and lay their eggs along the edge of the puddle, waiting for the next insect to continue the cycle.

Not all worms are parasites. Annelids include the familiar earth worm found in gardens throughout North America. Found at any local bait and tackle store, these worms have helped fishermen catch fish around the country. Basically, earthworms are a muscular tube that burrows through the ground ingesting organic material and aerating the soil. A relative of earthworms are leeches, an ectoparasite, but not a harmful one. Fortunately, they don't spread diseases or cause health problems like other parasites. In fact, they have been studied for potential medical benefits.

Nematodes are the round worms with at least 25,000 species, half of which are parasitic, including heart worms. They are found in almost every ecosystem and in great numbers. It has been estimated that they may account for 8-10% of all the animals on Earth.

Chapter 10

Vertebrate Diversity

Bigger, Faster, Smarter
The vertebrates rule the world

Introduction

It can be difficult to choose the most successful lineage of animals on the planet. But, a case could be made for the vertebrates. With 64,000 species, vertebrates include the largest animals, the fastest animals on the land, in the air, and in the water. And of course, they include humans, clearly the most intelligent animal to have ever inhabited the Earth.

Modern vertebrate diversity is the result of more than 500 million years of evolution, extinction, and geological forces. Vertebrates have successfully colonized every continent and are found in every ocean. They range in size from a tiny minnow less than a third of an inch, to the blue whale, reaching over 100 feet. Although vertebrates got their start in marine environments, one lineage, the tetrapods successfully colonized the land, and taken to the air three separate times. They also returned to the oceans as marine mammals, sea snakes, and sea turtles, along with several types of extinct reptiles from the Mesozoic Era. Importantly, vertebrates evolved endothermy (warm blooded), the ability to generate their own body heat allowing them to be active in any environment, regardless of the temperature.

Since the first simple chordates, the evolution of an internal skeleton, hard teeth, jaws, paired appendages, swim bladders, and lungs helps to explain their success as these features have been continuously modified and re-purposed over time. For example, fins evolved into limbs, scales into feathers and hair. Swim bladders used by fish to regulate buoyancy evolved Into lungs for breathing on land. Jaws used to improve oxygen absorption in water also evolved sharp teeth to catch prey. Two-chambered hearts evolved into three chambers, then four chambers, maximizing oxygen delivery to increasingly metabolically demanding muscles and larger animals. An internal skeleton providing muscle attachment for rapid swimming was strengthened to live on land. And so it goes, the vertebrates continually evolved and diversified into an incredibly successful group.

A Brief History of Vertebrate Evolution

The vertebrate story possibly began with a small worm-like animal called *Pikaia* swimming in the ancient Cambrian oceans over 500 million years ago. Despite its simple structure, *Pikaia* had many of the features still found in vertebrates today that unite this entire phylum to a single common ancestor. Based on fossil evidence, molecular evidence, and shared characteristics, it is clear that all modern vertebrates descended from a single common ancestor.

All vertebrates possess pharyngeal slits, a dorsal nerve cord, a notochord, a muscular post-anal tail, and mineralized tissues including teeth, cartilage, and bones. Although, I should point out that some of these features have become highly modified. But most importantly, vertebrates possess a vertebral column made of vertebrae to protect the nerve cord and provide rigid support for muscle attachment. In primitive chordates, they lacked a vertebrae and the notochord was a stiff cartilaginous rod running parallel to the nerve cord. The main purpose of the notochord was to support the body and provide a structure for muscle attachment helping it swim by undulating its tail. By having a stiff rod running the length of the animal, its body would not shrink when the muscles contracted, allowing it to quickly swim through the water.

Starting about 450 million years ago in the Ordovician, a wave of major evolutionary innovations began to take place with the origins of jaws, teeth, and paired appendages (pectoral and pelvic fins). Natural selection is like "tinkering". Over time, the ancestral features of vertebrates became increasingly modified, as vertebrates have continuously evolved to exploit new resources in a constantly changing world.

A great example of evolutionary tinkering occurred in the pharyngeal (gill) slits found in sessile tunicates where they are used for filter feeding. In the earliest fish, the pharyngeal slits were first modified into gills to improve gas exchange, getting more oxygen to metabolically demanding muscles. With the evolution of gills for gas exchange, our ancestors went from stationary filter feeders to active predators. The front gill slits later evolved into jaws that at first were used to increase water flow over the gills, allowing for even more physical activity due to greater oxygen flow. But, sometimes serendipity happens in evolution; the same jaws used to improve gas exchange in the water also allowed fish to bite harder, allowing them to take a larger variety of prey.

Fish with jaws are called Gnathostomes, which literally means "jawed mouth". With sharp teeth and paired fins, they were, and have remained formidable predators. In fact, the rise of jawed fish was probably responsible for the extinction of the more primitive Ostracoderms and other jawless fish that once dominated the ancient Devonian oceans. The fact that Ostracoderms lacked jaws but had armor made of teeth-like structures, indicate they were potentially filled the transition between jawless fish and jawed fish. Teeth and other mineralized tissues, such as bones and cartilage, are made of a mineral you probably never heard of called hydroxyapatite. Comprised of calcium and phosphorous it is much tougher than calcium carbonate, used by mollusks and reef-building corals.

During the Ordovician, jawed fish split into two lineages, the cartilaginous fish and the bony fish. The cartilaginous fish include the sharks, skates, and rays. Modern sharks have changed little from their ancestors, which first appeared 360 million years ago. When it comes to diversity, the bony fish are the clear winners. Their success is likely tied to the evolution of a gas-filled swim bladder allowing the fish to control its buoyancy in water. A swim bladder saved the fish energy from constantly having to swim and freed the fins allowing for greater maneuverability.

Shortly after their arrival at the end of the Silurian, bony fish split into two major lineages, the lobe-finned fish and the ray-finned fish. Today, when we think of fish, it's the ray-finned fish that come to mind. About 30,000 species of ray-finned fish inhabit every part of the oceans from crushing depths more than 10 miles below the surface, to the near freezing waters of the polar regions. Not to mention, you can find fish in lakes and rivers throughout the world.

By the end of the Devonian, 359 million years ago, fish had become quite diverse, with some calling it the "Age of Fishes". About 375 million years ago, a particular species of lobe-finned fish evolved into the first tetrapods (tetra=4, pods=foot), and the colonization of land by vertebrates had begun. Once again, serendipity and evolutionary tinkering helped tetrapods thrive on land. The swim bladder had evolved into lungs and was further modified for a life on land. The pectoral and pelvic fins evolved into the four limbs of modern tetrapods. Even the ancient bone arrangement of one bone, two bones, and lots of bones, has been preserved in all the major lineages of tetrapods. The pharyngeal gill slits were no longer needed for gills and evolved into parts of the jaws, necks, and ears in mammals. Mineralized bones used for muscle attachment in fish became larger and strengthened to support vertebrates on land. The two chambered heart of fish evolved into three and then four chambers allowing tetrapods to grow large on land.

Since their invasion of land 375 million years ago, tetrapods have diversified into modern amphibians, reptiles, birds, and mammals. What that means, is that we are descended from a fish ancestor. This may seem like a far-fetched claim, but the evidence is recorded in the fossil record, seen in our own anatomy, and written in our DNA.

Sharks, Skates, and Rays
How Nature got it Right

The first cartilaginous fish inhabited the Earth over four hundred million years ago. Today, there are close to 900 living species, which mostly include the sharks and rays. The ancient sharks, similar to modern sharks, appeared about 420 million years ago and have changed very little since, it's as if nature got sharks right from the start. Although, the first truly modern sharks date back to the Jurassic about 144 million years ago. The major difference being in the position of the mouth; modern sharks possess a much more pronounced rostrum, possibly an adaptation for small electro-receptors used to detect their prey. The ancestors of rays, easily identified by their flattened body, diverged from the sharks about 200 million years ago.

Despite their lack of diversity compared to the bony fish, sharks and rays are an incredibly successful group. First, they represent one of the earliest radiations of jawed fish over 420 million years ago. Then, about 360 million years ago, they underwent another round of rapid radiation in diversity, which may have led to the demise of the once successful placoderms. The placoderms were an ancient relative of sharks with hard armored coverings that once dominated the Devonian oceans during the Age of Fishes. Today, cartilaginous fish remain successful, as some of the most common apex predator in the oceans. Although over-fishing has drastically reduced their populations, making many species vulnerable to extinction. Unfortunately, they reproduce slowly making it difficult for populations to recover from over-fishing. Not only do most sharks reproduce slowly, some species live a long time. The Greenland shark (*Somniosus microcephalus*) may live to 400 years old, making it the oldest vertebrate animal.

Cartilaginous fish get their name from the lack calcified or mineralized bones, so complete fossils are rare. Surprisingly, recent DNA evidence indicates that their lack of hard mineralized bones may actually be a derived trait. The implication is that their ancestors forsook mineralized bone, which may have provided them greater agility. However, their teeth are made of the hard mineral hydroxyapatite, the same mineral used by all jawed fish, including ourselves, to make teeth. Shark teeth are abundant in the fossil record and often the only evidence we have of an extinct species.

It's well known that sharks have powerful jaws filled with rows of sharp teeth where their size and shape determine the type of their prey. Shark constantly replace their teeth and a single shark can shed upwards of 20,000 teeth in a lifetime, a good reason why they are common in the fossil record. What you may not know is that the scales of sharks and all cartilaginous fish are miniature versions of teeth called placoid scales. It takes only small changes in a shark's DNA to change the shape of a tooth into a smaller placoid scale. It's just a matter of changing its shape and size, everything else remains the same when it comes to the structure of a shark tooth. In fact, their teeth and scales have dentine and enamel, just like our teeth.

Sharks and rays lack a swim bladder; therefore, they must constantly swim, or they will sink to the bottom of the ocean. Additionally, watched sharks swim with their mouths open, showing off their sharp teeth? Like other fish, sharks get oxygen from water flowing over their gills, so they swim with their mouth open to force water across their gills. Although they lack a swim bladder, large pectoral fins provide additional lift while they swim. If a shark gets tangled in a net or caught on a long-line, then it will drown because it cannot move water across its gills. Some sharks, including nurse sharks, and almost all rays actively pump water across their gills through their mouths, or through small openings near the eyes called spiracles.

We don't know much about shark reproduction in terms of where they reproduce, how often they reproduce, and how potential mates find each other. We do know that sharks and rays either lay eggs while others give live birth to small pups. Live birth in sharks is different from mammals. In sharks that give live birth, the eggs remain inside the female until they hatch. Sometimes, the baby sharks, remain inside the mother growing larger to improve their chances of survival when they go off on their own. Sometimes the pups may eat their siblings with only the largest ones surviving to leave the mom. Unlike most ray-finned fish, sharks reproduce internally, a derived trait compared to their ancestors.

In recent years, scientists attached satellite trackers to sharks, which revealed new information about their movements and habits. Several interesting discoveries relate to their world-wide movements. For example, one shark swam from the California coast to Hawaii, while another swam from South Africa to Australia, sometimes swimming in total darkness more than 2500 feet below the surface for thousands of miles. No one is quite sure why they make such long migrations or how they navigate the ocean's depths.

Compared to us, sharks have extra-sensory perception (ESP), or senses in addition to sight, sound, or smell. On their head is a network of electroreceptors comprised of small gel-filled openings filled called the ampullae of Lorenzeni, which detects the weak electric fields of animals. Every animal produces a weak electric field due to the action of their nervous system and muscles. Hammerheads have taken their ability to detect electrical fields to an extreme as their head has evolved into a broad hammer shape to provide a wider area for the ampullae of Lorenzeni to detect small crustaceans buried in the bottom of the ocean.

Obviously, sharks are successful predators. They possess several additional adaptations to help them home in and catch their prey. The Great White Shark ranks among the most well-known and feared predators in the world and for good reasons. Like other sharks, they use the ampullae of Lorenzeni to detect the beating heart of their prey. In addition to detecting electrical fields, they can smell tiny amounts of blood in the water so they can find wounded prey. Great whites are dark on the bottom and light on the top, a color pattern known as counter shading, making them difficult to detect either above or below them. And of course, their large powerful jaws full of sharp teeth tear flesh apart. It's easy to see why the Great White is an apex predator, at the top of the food chain. They can reach lengths upwards of 20 feet, however that pales in comparison to their extinct relative *Megalodon*. Some estimates suggest that *Megalodon* reached lengths in excess of 50 feet and most likely preyed on whales. About 2.9 million years ago, they went extinct for unknown reasons.

Not all sharks are top predators, in fact, the largest living shark, the Whale Shark often exceeds 50 feet in length. These giant filter feeders swimming through the ocean collecting small shrimp and fish. Part of the reason why they reach such large sizes is because they feed closer to the base of the food chain where more prey and energy are available.

The skates and rays represent the other major lineage of cartilaginous fish with their flattened bodies and gills located on their underside. Rays are kite-shaped with long tails, and the stingrays possess one or two stingers on their tails. Similar to some sharks, rays give birth to live young, but skates lay eggs called a mermaid's purse, which often washed up on sandy beaches. Reaching 27 feet across, the manta ray is the largest ray, despite their large size, they can jump out of the water. Manta rays are harmless filter feeders, often observed by divers.

Perhaps one of the most interesting rays is the sawfish, which bears little resemblance to other rays. First, their elongated body resembles a more "shark-like" appearance. Despite superficially looking like a shark, their gills are located on their underside, like all other skates and rays. However, the most unique feature of Sawfish is an elongated saw-like rostrum with numerous ampullae of Lorenzeni, the electro-sensitive pores allowing sawfish to detect prey hiding under the sea floor. When they find their prey, they slash at it with their saw-like rostrum.

Bony Fish: An Evolutionary Success Story

Bony fish first appeared over 420 million years ago, about the same time as the cartilaginous fish. The rise of bony fish was an evolutionary leap for several reasons. They get their name because they form true bone made from a specialized cell called an osteocyte. Bone is different from cartilage in several ways, importantly bone is a dynamic tissue capable of growth and repair. Their true success, however, may have been caused by the evolution of the swim bladder, which may have started as a primitive lung for gas exchange with the atmosphere. You read that right, lungs evolved in fish before terrestrial vertebrates. Some scientists have hypothesized that fish evolved lungs because they lived in warm areas where dissolved oxygen levels were temporarily depleted, or they simply required more oxygen for their increasingly active life styles. They think that the lungs were later modified into a swim bladder allowing bony fish to control their buoyancy.

The evolution of swim bladders is another example of evolutionary serendipity because it helped fish in two important ways. First, it saved the fish energy from having to constantly swim. Recall that sharks lack a swim bladder and sink when they stop swimming. Second, it freed the fins from being used to maintain lift in the water and allowed them to be modified for maneuverability. Together, these modifications helped to make bony fish quite successful.

There are two major linages of bony fish, the ray-finned fishes and the lobe-finned fishes, which include the tetrapods. You are most familiar with the ray-finned fish, including clown fish made famous by the Disney film *Finding Nemo*, along with popular game fish such as tuna, salmon, snapper, grouper, and dauphin fish (popularly called Mahi Mahi to make it more desirable at market and fetch a higher price).

Diversity of lobe-finned fish peaked in the Devonian and declined by 365 million years ago. Today, there are only 9 known species of lobe-finned fish still living. However, their ancestors did give rise to the tetrapods, which include about 30,000 living species, a rate of diversity similar to the ray-finned fishes. In contrast, ray-finned fish remained largely confined to watery environments, never fully making the transition to land. They can be found just about anywhere with permanent water, with the exception of extremely hot, saline, or acidic environments.

If you're like me, I'm interested in the biggest and the fastest animals. The fastest fish in the ocean are the sailfish (*Istiophorus*), which can reach top speeds of 68 miles per hour in water! The Bluefin Tuna grow over 10 feet in length and weigh over a thousand pounds, yet it can reach speeds close to 50 miles per hour. Until recently, it was generally assumed that all fish were ectothermic (ecto=out, therm=heat), which means their body temperature depends on the temperature of their environment. We commonly refer to ectotherms as cold-blooded, but that can be misleading as many ectotherms are most active at the same temperature as a mammal or a bird.

In true scientific fashion, a scientist caught a small tuna and was amazed at how fast it could move its muscles considering it was caught in cooler water. Muscles are made or proteins that contract faster in warmer temperatures. Upon making the observation of the fast speed of the tuna's muscles, he measured the temperature of the fish to find that its muscles were 10 degrees above the temperature of the water. This one observation changed decades of thinking about fish and also showed the importance of curiosity, making observations, asking questions, and experimentally testing our assumptions.

More fish with ESP? In Africa, you can find elephant fish in the family Mormyridae sporting a long "snout" used to generate weak electric fields. In addition to detecting prey, they also use the electric fields to "see" their surroundings. The snout and body possess sense organs similar to the ampullae of Lorenzeni in sharks. Some evidence suggest that elephant fish communicate through their electric fields. Surprisingly, the brain to body ratio of elephant fish rivals human, perhaps to handle their sensory inputs. Electric eels and electric catfish also produce electric fields delivering a jolt of several hundred volts used to stun their prey!

When we think of venomous animals, rattlesnakes usually come to mind, but there are about 1200 species of venomous fish! There's a difference between a venomous and a poisonous animal; a venomous animal injects a toxin into an animal either through fangs (modified teeth), fins, or stingers, whereas a poisonous animal is toxic if you ingest it or absorb the toxin through the skin.

The most popular venomous fish are the lionfish (*Pterois*), a popular aquarium species. Their long spines inject a neurotoxin causing severe pain and in rare cases lead to death. As a result, they have few natural predators, a major reason why their populations have exploded in the Atlantic making them an invasive species. Fins are not the only venomous part of fish. One small marine fish found on tropical reefs, called the fang-toothed blenny (*Omobranchus ferox*), uses fangs on their lower jaw, for self-defense.

Ray-finned fish come in all shapes and sizes and you can tell a lot about a fish just by looking at its shape. Torpedo shaped tuna maintain high speeds for hours in the open ocean. Whereas, the leafy sea dragon (*Phycodurus eques*) is the exact opposite, its fins resemble sea weed to avoid detection by both predators and prey. Flounder are flattened to hide under sandy bottoms to ambush their prey. With their small bodies and enormous mouths full of long sharp teeth, the scariest looking fish live in the deep sea. There is not a lot of food in the deep sea, a meal might come along only rarely. To deal with the scarcity of food, many deep-sea fish evolved huge mouths to catch large prey so they can go long times between eating. Some of these fish formed symbiotic relationships with bioluminescent bacteria to attract prey.

With a long evolutionary history, ray-finned fish evolved some rather unusual mating habits. For example, the males of deep-sea anglerfish are barely more than degenerate sperm donors. It is hard to find potential mates in the deep sea, so the male permanently latches to the female's skin for the rest of his life, fertilizing the female's eggs when needed. The bonding is so complete that even their tissues fuse together to the point where the blood vessels between them merge providing the male with all the nutrients and energy he needs.

We typically assume that females watch over the babies, but this is not always the case. In sea horses, the male carries his babies in a pouch located on his abdomen. Strangely, the males of some fish brood their babies in their mouths, protecting them from predation.

External reproduction is the ancestral trait for vertebrates, but internal reproduction evolved more than once in the vertebrates. The common guppy found at pet stores and mosquitofish, common throughout the U.S., reproduce internally giving birth to live young.

In fact, the least killifish (*Heterandria formosa*) of the southeastern US is the smallest live-bearing vertebrate in the world. While the sex in mammals is determined by genetics, many species of fish change sex based on who's in their environment (I'll discuss this more in Chapter 6).

Lobe-finned Fish Gave Rise Tetrapods

For some, it's difficult to believe we evolved from fish, I mean we look nothing like a fish. But, the distant relative of all tetrapods, animals with four limbs including amphibians, reptiles, birds, and mammals, evolved from a lobe-finned fish living about 375 million years ago. We aren't exactly sure what drove lobe-finned fish to colonize the land. Although, several good hypotheses have been proposed to explain why they transitioned from water to land.

The first hypothesis invokes competition, a force of nature that constantly drives the evolution of new species and innovations. Recall that the Devonian was the Age of Fishes, a time when fish were morphologically quite diverse and likely fueling intense competition. Any species that evolved to live on land would experience less competition with other fish. The second hypothesis proposes that the move to land was to exploit new resources. When fish were making the transition to land, there were no other terrestrial vertebrates, but large forests were present, and arthropods had been there for millions of years. By moving to land, vertebrates could exploit new resources available that weren't available to them. Most likely, the transition to land was probably driven by a combination of less competition from other fish and the move provided opportunities to exploit new resources.

During the Devonian from 419 to 359 million years ago, the lobe-finned fish were in their heyday of diversity. Of all the lineages of fishes, they were the best suited to make the transition to land. So, how does a fish evolve into a tetrapod? A common theme in evolution is that natural selection acts to modify features already present and re-purpose them for new uses. The re-purposing and modification of lungs, paired limbs, and a bony skeleton played a major role in the evolution of lobe-finned into terrestrial tetrapods.

About 380 million years ago, the transition to land likely began when a lobe-finned fish started spending part of their time at the water's edge and possibly some time on land. Their primitive lungs helped them absorb oxygen from the atmosphere. Recall that fossils suggest lungs first evolved in fish living in slow moving or stagnant water. During times of warm temperatures, the water lacked sufficient oxygen for fish to be active or even survive. To make up for their oxygen debt, these fish evolved lungs to gulp air to survive. The end result is that by having a lung already present, it made the transition to land easier because they already had means to breathe air. Although, it's important to state that this was serendipity, evolution is not evolving towards a future goal.

Both ray-finned and lobe-finned fish have paired appendages, two pectoral fins and two pelvic fins. However, the fins of the lobe-finned fish are connected to the body by a single bone attached to muscles. Out in the fin, the single bone is attached to other bones. This arrangement helped to support their body on land. As they continued to evolve to life on land, their fins eventually evolved the universal configuration of one bone, two bone, lots of smaller bones, including a wrist allowing them to easily walk on land. Even today, the bones in our own limbs still follow the same pattern of all other tetrapods of one bone, two bones, lots of bones.

Don't forget, Lobe-finned fish are also a bony fish, meaning they have a mineralized skeleton. In watery environments, the skeleton would provide sites for muscle attachment allowing fish to swim with much more power and speed. The bony skeleton was another important adaptation in fish that was helpful in the transition to land. Once on land, the first tetrapods felt the force of gravity, which was not important in water. Fortunately, the bony skeleton was easily modified to become larger and thicker and therefore capable of supporting their bodies on land. It's almost as if the lobe-finned fish with their lungs, lobed fins, and bony skeleton were pre-adapted to living on land.

Today, only 9 species of lobe-finned fish survive, including the Coelacanth found in the Indian Ocean and a few species of lungfish found in freshwater habitats in Africa and Australia. Despite the lack of modern diversity in lobe-finned fish, their legacy as tetrapods continues today, rivaling their ray-finned cousins.

Amphibians Were the First Tetrapods

Most tetrapods breath air through lungs (although, there are a few lungless salamanders). The first amphibians resembled a large salamander while retaining some fish-like qualities, such as scales. Reminiscent of their fish origins, these animals returned to the water where they reproduced externally. Although, the first fully formed amphibians appear in the fossil record about 370 million years ago, modern amphibians, which include three lineages, the frogs and toads, salamanders, and the worm-like caecilians first appear in the fossil record about 240 million years ago. This was about the same time that dinosaurs appeared and began to dominate the land. Today, about 7,000 species of amphibians, 90% are frogs and toads inhabit the Earth.

Modern amphibians still reproduce through external reproduction. However, as you can imagine, they are quite diverse in the details of where and how they reproduce. While many amphibians continue to reproduce in water, others do not.

Laying eggs in permanent streams or water bodies poses a problem, they face predation from fish. Therefore, some frogs have come up with unique adaptations to prevent their eggs or tadpoles from being eaten. Glass frogs lay their eggs on leaves over a stream where they will guard them. When it rains, the eggs will wash into the stream and develop into small frogs. A poison dart frog in South America called *Colostethus* will carry their tadpoles on their backs. Other poison dart frogs will lay their eggs in bromeliads, a common air plant found growing on trees high up in the canopy. Because of the abundant rain, water pools in their folded leaves, providing a safe place for their eggs. Surprisingly, frogs in the genus *Craugastor* and the Coqui frogs have direct development, they lay eggs in the leaf litter and the eggs develop directly into small froglets, allowing these frogs to skip laying their eggs in water altogether.

In the desert southwest, spadefoot toads survive where water is scarce. During the summer monsoon, you can slowly drive country highways at night (commonly known as road cruising), and if you're lucky, you will find spadefoot toads sitting in the middle of the road. Spadefoot toads get their name from a small spade on their hind legs that help them burrow into the ground where they spend most of their lives.

With little to no permanent water anywhere in sight, these small toads remain burrowed underground until the arrival of the monsoonal rains in early July. In low lying areas, temporary ponds form, holding water a few weeks to a few months, depending on the amount of rain and their location. With the first rains, spadefoot toads emerge from their burrows, mate, and lay their eggs. The tadpoles quickly morph into small adults before the ponds dry up. At the end of the monsoonal rains, they will burrow back into the ground and wait for the next monsoon season.

The Darwin's frog (*Rhinoderma darwinii*) of South America uses perhaps the most bizarre method to protect their young. After mating, the male swallows the eggs and holds them in his vocal sac. Once the tadpoles grow into small frogs, the father will cough them up one at a time. He produces a nutrient rich liquid, similar to mammalian milk to help nourish his progeny. In Australia, there were two species of gastric brooding frogs in the genus *Rheobatrachus*. The females would incubate the larvae in her stomach until they were older, and similar to the male Darwin's frog, she would cough them up when they were small frogs. Unfortunately, both species of gastric brooding frogs are most likely extinct due to habitat loss and the Chytrid fungus. However, Australian scientists began the "Lazarus Project" to bring back the species by cloning them.

Frogs and toads have a unique body plan, their hind legs and pelvis region are specialized for jumping or hopping. In North America, the green tree frog (*Hyla cinerea*) can jump upwards of 150 times its body length, second only to the flea. Not every frog can jump so far, and toads merely hop, which makes them an easy target for predators. To protect themselves, some frogs and toads produce toxins in special glands or in their skin making them taste bad. For example, toads have a large warty-looking bump behind their eyes, called a parotid gland, which produces a neurotoxin called bufotoxin. It is potent enough to kill most animals interested in eating it. In Central and South America, small, brightly colored poison dart frogs in the family Dendrobatidae produce toxic substances called alkaloids.

Salamanders, which resembles the first tetrapods, date back to the middle Jurassic, about 170 million years. Although, modern salamanders lack scales and do not grow as large as their ancestors. Today, the Chinese giant salamander (*Andrias davidianus*), which can reach a length upwards of six feet holds the record for the largest living amphibian. Their large teeth causes harsh wounds when the males fight over their territories.

Lacking a pelvis and hind-limbs, the sirens (family: Sirenidae) appear eel-like. As adults, they possess external gills, lack eyelids, and burrow in the mud. Surprisingly, not all tetrapods have lungs. In North America the lungless salamanders in the family Plethodontidae, represents a lineage that has lost their lungs. Instead, they respire across their skin and through the tissue linings of their mouth. In the eastern United States, they can be quite numerous with population estimates of 1.8 billion in a single forest. Their small size and slow metabolism allow them to reach such high numbers. Salamanders have one very interesting ability of major interest to us, they can regenerate their limbs if they lose them.

Amniotes: Adaptations for Living on Land

Sometime around 312 million years ago as the world became drier and cooler, a group of tetrapods evolved several useful adaptations improving their ability to survive on the land by reducing their dependence on water. This new group of animals reproduced internally and placed the embryo in a protective, water-filled structure called the amniotic sac. The evolution of these two features, internal reproduction and the amniotic sac protected by a hard shell, allowed tetrapods to lay their eggs on land.

Shortly after the appearance of amniotes about 300 million years ago, they split into two lineages. One lineage evolved into reptiles, a diverse assemblage of animals including lizards and snakes, turtles, pterosaurs, ichthyosaurs, mosasaurs, plesiosaurs, crocodiles, dinosaurs, and birds. The other lineage of amniotes eventually evolved into modern day mammals. Amniotes are a lineage whose evolution was largely driven by the spread of dry environments. But, they have also repeatedly returned to both marine and freshwater habitats.

Reptiles
Vertebrates Conquering the Land

The ancestors of reptiles and mammals began to diverge a little over 300 million years ago. Despite the vast amount of time since their divergence, mammals and reptiles still reproduce internally and produce a water-filled amniotic sac to protect their embryos. Today, reptiles are a diverse lineage even though their heyday ended 65.5 million years ago with the extinction of most dinosaurs and their relatives. Today, we typically classify modern reptiles in two broad groups, the Archosaurs and the Lepidosaurs, although we aren't quite sure where turtles fit in. Ironically, Archosaurs means "ruling lizards", but they aren't true lizards. Living Archosaurs include birds and crocodilians, but there are many extinct forms including the dinosaurs and the flying pterosaurs. Lepidosaurs mean "scaled lizards" and include both snakes, lizards, and the primitive looking tuatara. Common extinct reptiles from the Mesozoic include marine forms such as ichthyosaurs which resembled modern dolphins, plesiosaurs, and the large mosasaurs like the one seen eating a shark in the movie *Jurassic World*.

Most reptiles are ectothermic, meaning their body temperature matches the environment, or they do not produce metabolic heat. While this lets them go longer between meals because of lower metabolism, ectothermy somewhat limits reptiles in their distribution into colder climates. Reptiles also have scales, an evolutionary adaption that prevents water loss across the skin. Additional features of their skulls, including the shape of the jaw, ear bones, and the number of openings in the skull, further separate them from mammals. Most reptiles have a three chambered heart, although birds and crocodiles have four a chambered heart that prevents the mixing of oxygenated and deoxygenated blood.

Turtles

With their hard shells, turtles form one of the most recognizable groups of reptiles. In fact, the uniqueness of turtles makes it hard to understand their relationship to other reptiles. Just how and when did turtles evolve their shells? Based on fossil records, their origins date back to the Jurassic Period 240 million years ago when they last shared an ancestor with other reptiles. That makes them an older group than dinosaurs or crocodiles. Today, about 327 species inhabit the Earth in a wide range of environments from the open ocean to deserts. Turtles live a long time, with giant tortoises living more than 200 years! Unfortunately, habitat loss and over harvesting are driving many species to extinction.

Many turtles have a close association with water. Freshwater turtles can be seen basking in the sun along the banks of streams and rivers. Sea turtles spend their entire lives at sea where only the females return to lay their eggs on sandy beaches. Living upwards of 80 years, the Leatherback is the largest turtle. Its evolutionary roots date back over 100 million years ago to a time when the dinosaurs dominated the land. These turtles can dive down to 4800 feet and hold their breath for 85 minutes. Leatherback sea turtles are found throughout the world's oceans eating jellyfish. Unlike other reptiles, they exhibit a small degree of endothermy. Because of their large size, they maintain body temperatures above their environment, enabling them to survive in colder waters.

The remote ancestor to leatherback sea turtles and all other sea turtles once lived on land. Today, sea turtles make long migrations, with some individuals swimming over 3700 miles one way to lay eggs on the same beach where they hatched. Surprisingly, continental drift may explain why some make such long migrations. A hundred million years ago, their ancestors didn't have to swim as far, but with the continents moving apart at the rate of an inch per year, that distance has become quite large over time. So each year, they swim a little further to lay their eggs.

Lepidosaurs
The Scaled Reptiles

Lizards, snakes, and the Tuatara belong to the Lepidosaurs (lepido = scaled saur = lizard). Despite not having legs, snakes are closely related to lizards. In fact, the loss of limbs in lizards has occurred multiple times. Worldwide, about 3,400 species of snakes live in a variety of habitats from sparsely vegetated deserts, to dense rainforests, and some even live in the open oceans. In some places, snakes can be quite common, but their cryptic and reclusive nature makes it difficult to estimate their populations.

Snakes use several unique senses to find and catch their prey in addition to eyesight. First, they flick their tongues to detect odors in the air and to taste the ground. Pit vipers in the Americas use small pits below the eye to detect heat.

When it comes to catching and killing their prey, snakes use various tactics, including constriction and venom. A constrictor will quickly grab their prey and then wrap around them multiple times. With each breath, the snake tightens down, eventually suffocating it. Snakes have modified jaws allowing them to swallow prey that are bigger around than they are. Contrary to popular belief, the jaws are not unhinged, they just aren't rigidly attached to the skull. Ligaments connect the lower jaw bone allowing each side to move independently, unlike our lower jaw where both sides move together.

People fear snakes, most likely because some are highly venomous capable of causing a lot of pain at the very least. Venomous snakes fall into one of two families, the Elapids and the Vipers. Found on most continents, the Elapids include cobras and coral snakes. In the U.S., the coral snake has a very potent neurotoxic venom that can cause respiratory failure within a few hours. Vipers are found in the western and eastern hemisphere. In the U.S., they include the pit vipers, most commonly known as rattlesnakes (*Crotalus*), copperheads (*Agkistrodon contortix*), and water moccasins (*Agkistrodon piscivorus*). Their venom can be neurotoxic or cytotoxic depending on the viper species and the type of prey. Rattlesnakes get their name from a rattle formed by a series of keratinized interlocking scales that rattle when shaken. Rattlesnakes retain their eggs inside until they hatch, giving birth to live young.

Lizards are closely related to snakes and can be quite common in some habitats. As of 2015, about 6,145 species inhabit all the continents except Antarctica. In the Western Hemisphere, the genus *Anolis*, has over 400 known species, making it the largest genera of amniotes in the world. Known for their dewlaps, this colorful patch of skin near the throat is extended by the males to attract females or establish territories. The colors and size of dewlaps differs between species. Some anoles, including the Carolina anole in the south eastern U.S., change their skin color to match their surroundings. Recent expeditions to the tropics, led by Dr. Steve Poe from the University of New Mexico, continue to discover new species of Anoles.

The geckos represent another successful lineage of lizards. To put their success in perspective, of the 6145-known species of lizards, 1500 are geckos. Because most geckos are nocturnal, their eyes are 350 times more sensitive to light than ours. In addition to light-sensitive eyes, their toepads allow them cling to variety of surfaces, and they have been observed walking across ceilings. Fortunately, some species of geckos do well in human environments, especially at night on walls with lights that attract insects.

The largest lizard in the world is the Komodo Dragon, a type of monitor growing up to 9 feet long and weighing about 150 pounds. It's unclear why this lizard is so large, but it may be relict of an older group of large lizards that went extinct about 15,000 years ago. Komodo dragons ambush their prey, taking invertebrates, birds, and mammals, including small deer. Interestingly, Komodo Dragons have been known to hunt together, a unique behavior among reptiles.

Archosaurs
The Ruling Lizards

It may seem odd to include dinosaurs, birds, and crocodiles into one lineage. After all, birds and crocodiles don't seem to have much in common. Feathers, wings, and bills lacking teeth make any bird instantly recognizable. Alligators, on the other hand, are large and sprawling, with powerful jaws full of teeth. But, if you look closer, they both have scales and four-chambered hearts, traits inherited from a common ancestor. Additionally, birds and alligators share several behaviors; they make a nest to lay their eggs and "sing" to their mates. Because both groups share these behaviors, it is generally accepted that dinosaurs also made nests and sang to their mates, once again we assume this based on common ancestry.

About 24 species of crocodilians remain today and many are imperiled. A nostril on top of their head allows them to breathe while remaining mostly under water. Crocodilians ambush their prey by quickly launching themselves out of the water, or rapidly snapping their mouths shut. Once they catch their prey, crocodiles will drown it by repeatedly rolling in the water. The Saltwater Crocodile (*Crocodylus porosus*) is the largest living reptile, males have been known to reach 20 feet and weigh over 2,000 pounds. By exerting 3,690 pounds per square foot, they have the largest bite force of any terrestrial vertebrate and rival adult Great White Sharks.

For nearly 165 million years of the Mesozoic Era, the dinosaurs dominated the land. So far, paleontologists have identified about 700 species of dinosaurs of diverse in sizes and habits. Based on fossils, the smallest dinosaur, *Microraptor* barely over a foot long to the massive *Titanosaurs*. These large herbivorous dinosaurs are estimated to have been 130 feet long and weighing up to 90 tons. The first dinosaurs evolved in the Triassic about 230 million years ago and walked on two legs. Some were carnivores and others were herbivores.

Whether dinosaurs were active endothermic animals is still up for debate. Although, some evidence indicates they were fairly active animals. By comparing the traits of their closest living relatives, the birds and crocodiles, we assume that dinosaurs had a 4-chambered heart, an important adaptation for large active animals. In recent years, new fossils have shown that dinosaurs potentially evolved feathers long before the first birds evolved.

We aren't quite sure why dinosaurs evolved feathers, but it probably wasn't for flight, that would come later. It is thought that feathers were used for insulation in colder temperatures, or used in displays to attract mates. Today, birds used feathers for both those functions today, in addition to flight. Also, modern birds generate their own body heat, making them endothermic, a necessary condition for powered flight. Endothermy, feathers, and a four chambered heart provide good evidence supporting that at least some dinosaurs were endothermic with active lifestyles. Although, the case is not closed on our understanding of dinosaur physiology, paleontologists are still actively searching for additional evidence in the fossil record.

In a recurring theme of evolution, a feature evolves for one purpose and then gets re-purposed for new uses. Evolving first in dinosaurs, feathers were likely used for insulation or to attract mates, but not flight. Starting about 145 million years ago, feathers were modified for flight.

Today, birds use feathers for all three purposes, flight, insulation, and to attract mates.

Although, most people currently accept that birds evolved from theropod dinosaurs; however, not everyone agrees, and recent analyses suggest that birds may have a different origin. Currently all evidence strongly supports their archosaur lineage, but their relationship to dinosaurs is not as well-known as thought by many scientists. The disagreement shows that more work remains to be done and we should all be cautious about supporting popular theories, accepting them without sufficient evidence, or ignoring new evidence that contradicts our previous beliefs.

With 10,000 species of birds today, their diversity is simply amazing, including their sizes, songs, and feathers. Not much bigger than an eraser at the end of your pencil, the tiny Bumblebee Hummingbird, weighing less than an ounce is the smallest bird. In fact, their size is about the lower limit for endotherms. Because of their small size, hummingbirds rapidly lose heat. To keep up with the rapid heat loss, their metabolism is so high they could die within a few hours of not eating. To get around this problem at night when they do not forage, hummingbirds go into torpor by lowering their body temperature to conserve energy. At the other end of the size spectrum is the Ostrich, a large flightless bird of an ancient lineage. They can stand almost 9 feet in height and weigh up to 200 pounds, and have been clocked running at 50 mph.

Although most birds can fly, there is a wide range in their flying abilities. The Greater Roadrunner of the American Southwest rarely flies, but you can find them running through your neighborhood. When they do fly, it is mostly to avoid predators and they don't fly very far. The fastest birds are the Peregrine Falcons, which been clocked at speeds of 240 mph while dive-bombing their prey. Humming birds have the fastest wing beat, about 80 beats per second and they are the only bird that can fly backwards. Flying across the open ocean, the Wandering Albatross used very long and narrow wings built for soaring, riding air currents for days at time. Although they are graceful in the air, they need a running start to fly.

In addition to flight, birds are known for their songs. While most birds make some sound, even if it is little more than grunt, the song birds take it to the next level. Songbirds account for about half of all bird diversity. The vocal cord in birds is called a syrinx and is analogous to our larynx. In some song birds, the syrinx is split in two, allowing birds to make two sounds at once, which can be heard in Wood Thrushes and Hermit Thrushes. These brownish birds are commonly heard singing their melancholy sounds in forest throughout North America in the summer.

The bills and feet of birds are also adapted to very different lifestyles. Web feet evolved independently in penguins, gulls, pelicans, and ducks, all used for swimming on and in water. Likewise, the bills of these birds are also quite diverse. Pelicans use a pouch-like beak to scoop in water and food at the same time. Penguins have much smaller bills for catching fish directly. And ducks have wide bills for filtering plant material from the water. Although, some ducks have a long and narrow serrated bill to help them catch fish.

Songbirds are also quite varied in their diets and thus the shape of their bills. Finches and cardinals use thick bills for cracking seeds, whereas warblers use much smaller and thinner bills for catching insects. Often seen on sandy beaches and mudflats, sandpipers use tubular bills to probe the soil for small invertebrates. Their bills of varying lengths reduce competition between the species by foraging at different depths and habitats.

All birds lay eggs, not a single species gives live birth. Flight is an incredibly metabolically demanding activity, and birds have evolved numerous adaptations to accommodate their lifestyle. They are endothermic, generating their own body heat, which is necessary for fast muscle contractions. Their bones are hollow and fused together to conserve weight and provide for added strength. One really interesting adaptation is their unidirectional lungs, meaning air only goes one way through their lungs to maximize oxygen uptake.

Birds live a long time when compared to most mammals of similar size, and they rarely show signs of aging in the wild. In a remarkable series of photographs, a scientist caught the same seabird about once a decade over a period of 50 years. In each photograph, the scientist can be seen aging, but the bird always looked the same. To understand the aging process, scientist have begun to study birds to determine how they live so long.

Mammals
Bigger, Faster, and Smarter

Mammals share a common ancestor with the reptiles dating back 310 million years ago to the first amniotes. Shortly afterwards, amniotes split into two lineages, one became the mammals. Surprisingly, the ancestors to mammals dominated the land until the End Permian 251 million years ago, this was a massive extinction wiping out 90% of marine life and 60-70% of terrestrial life.

Twenty million years later, the dinosaurs had arrived and began to dominate the land, successfully out-competing our mammalian ancestors. But, the ancestor to mammals continued to evolve in the shadows of the dinosaurs, including traits allowing them to remain active at night, when temperatures were cooler and dinosaurs may not have been as active. To maintain high activity levels regardless of the temperature, mammals evolved endothermy where body heat is made through metabolism allowing mammals to be active regardless of the environment. For every adaptation allowing you to exploit a new resource, like endothermy to be active at night, there is a trade-off. In this case, mammals needed more food to maintain their high metabolism. Hair evolved to insulate them from heat loss. Additionally, mammals evolved an efficient four-chambered heart to increase their level of activity.

Early mammals also developed a keen sense of smell to help them find food at night, avoid predators, and find potential mates. By 205 million years ago, ancient mammals would be easily recognizable today. For the next 140 million years, most mammals remained small and nocturnal. Then, 65.5 million years ago, the dinosaurs went extinct, creating an opening for the mammals to diversify.

Modern mammals are a diverse group with a long evolutionary history. Mammals had been around for 140 million years when the dinosaurs went extinct, but by the beginning the Cenozoic Era 65.5 million years ago, they rapidly diversified to fill the ecological spaces previously held by dinosaurs. Although, for the first few million years after the dinosaur extinction, birds were also diversifying and competing with mammals for dominance. However, mammals eventually secured their dominance in terrestrial ecosystems, including the top predators.

As mammals diversified, some returned to the oceans, others took to the air, and some stayed in the trees only to come down in the last 5 million years and walk upright. The features mammals evolved during the Mesozoic Era, including endothermy, hair, and a four-chambered heart, were important factors priming mammals for their eventual success. Today, approximately 5400 species of mammals live on every continent except Antarctica, and they are found in the oceans, on the land, and in the air.

There are three major lineages of living mammals, monotremes, marsupials, and eutherians. Monotremes are an ancient egg-laying lineage including the Platypus and Echidnas of Australia. They are still considered mammals because they have hair and produce milk for their young. About 160 million years in the Jurassic, Eutherians split from the marsupials, forming a new mammal lineage where the fetus is carried in the uterus and connected to the mom by a placenta.

The Marsupials include about 334 extant mammals with the young being born at a very early age of development. Many species carry their young in a pouch. Basically, marsupials come in two major lineages, the American marsupials and the Australian Marsupials. Although, fossil and molecular evidence indicate that Marsupials may have reached Australia by way of Antarctica from South America over 50 million years ago during the break-up of the last remnants of Pangaea. Today, about 70% of all marsupials live in Australia, with another 100 species living mostly in South America. Marsupials include several well-known species including koalas, kangaroos, possums, opossums, bandicoots, and the Tasmanian devil.

Molecular and fossil evidence suggests that a single marsupial species diversified into the all the marsupials on the isolated Australian continent. Kangaroos, one of the best-known Australian marsupials, is the largest animal to move by hopping on large hind legs. Another well-known group of marsupials are the tree-dwelling Koalas who spend their time eating the leaves of eucalyptus trees. Interestingly, they have one of the smallest brains relative to their body size for any mammal. It takes up only about 60% of the brain case. Fluid fills the rest of the space to protect their brains when they fall. About the size of a small dog, the

Tasmanian devil is the largest carnivorous marsupial.

Eutherians, a name which means "new beast", forms the third major lineage of mammals with over 5,000 named species. They, like the reptiles, evolved flight and return to the oceans. Eutherian mammals range in size from tiny shrews weighing only a few grams to the Great Blue Whale weighing over 170 tons. They include many of the commonly encountered and well-known mammals including deer, elk, cows, bears, cats, dogs, rodents, bats, sloths, elephants, primates, and cetaceans along with strange looking scale-covered pangolins and armadillos. The two most diverse groups of mammals are the bats and rodents.

On an African safari, you get to see many mammals on the Serengeti plains. Many of these animals are ungulates a large group of even-toed mammals including cattle, pigs, giraffes, rhinoceroses, hippopotamuses, and wildebeest. In the U.S., we have our own ungulates, including mountain goats, bighorn sheep, elk, and deer. In the Serengeti, several species of large cats hunt many ungulates by ambushing them. To defend themselves, the ungulates, such as the wildebeest, form large herds to minimize their risk of being caught. Other mammals, such as water buffalo, became very large making it difficult to be preyed upon, while others remain small and agile to outrun the predators.

Ungulates belong to a group of animals that grow antlers or horns. Antlers are bony extensions of the skull grown by deer and elk, and are shed each year. Cows, goats, and sheep have horns, a permanent keratinized structure, which unlike horns does not get shed. Instead, they continually grow throughout the animal's life. Interestingly, even-toed mammals related to elk, deer, and hippos share a common ancestor with whales.

Cetaceans, commonly known as whales and dolphins include the largest animals, at least by weight, to have ever inhabited the Earth. The Blue Whale reaches lengths of 100 feet and weigh upwards of 191 tons or 382,000 lbs. To achieve such large size, they feed on small shrimp called krill which are at the base of the food chain. Humpback whales, known for their intricate songs sung by the males, will start a song, migrate thousands of miles, and then return to the same spot and pick up singing where they left off months earlier.

Whales share little resemblance to other mammals, but the

evidence of their relationship lies in the fact that their front fins maintain the same bone pattern of one bone, two bones, and lots of bones as other tetrapods. They also breathe air, are endothermic, grow hair, give birth to live young, and feed them milk. In some cases, up to 100 gallons per day for larger whales.

Not all cetaceans are giant filter feeders. The toothed whales include dolphins, orcas, and sperm whales, which are active predators. Dolphins form pods, sometimes with hundreds of individuals. The Bottlenose Dolphin is commonly encountered along the Gulf and Atlantic coasts in pods ranging from 10-30 members. They hunt for schools of fish as a group coordinating their efforts using echolocation. Dolphins have very large brains, with a brain to body size ratio similar to humans.

Mammals are excellent predators on the land, although none have achieved the same size as the largest dinosaur predators. Carnivores include well-known mammalian predators in the order Carnivora, which includes, wolves, cats, raccoon, bears, hyenas, otters, and badgers. These predators use different methods of catching their prey. The cats ambush their prey. Lions, the largest cats weighing up to 500 pounds, hunt in small groups called prides, allowing them to take on larger mammals such as wildebeest, zebras, and buffalo. While the smaller cheetahs, rely on speed. Sprinting to nearly 60mph, they outrun their prey slowing down before they overheat. Wolves and wild dogs will chase their prey over longer distances slowly wearing it out.

Bats are the second most diverse group of mammals in the world. One look at their wings and it becomes apparent they evolved flight independently of birds. Rather than using feathers, skin stretched between their digits forms the wings. Active at night, bats navigate and find prey with sonar by emitting high pitched sounds that bounce off their surroundings. As they approach their prey, they rapidly increase the number of "clicks" to home in on their prey. Bat flight may appear erratic; but in fact, it's quite precise as they scoop flying insects with their tail. Bats are voracious predators, consuming thousands of insects every night. Although, not every bat eats insects. In Central America, you can find the vampire bat, it uses modified teeth for cutting the skin of large mammals so it can drink their blood. The largest bats, the flying foxes, eat fruit and fly with a wingspan of 5 feet.

Rodents are the largest order of mammals with about 2270 known species accounting for about 40 percent of all mammals. They include many familiar animals, such as mice, rats, Guinea pigs, squirrels, chipmunks, prairie dogs, porcupines, and the capybara. You can find rodents on every continent, except Antarctica. They range in size from small mice to the 115-pound capybara of South America. Many species of rodents, such as prairie dogs, form large social groups and are vitally important to their ecosystems.

Sometime near the end of the dinosaur's reign, a small arboreal mammal evolved forward facing eyes and fingers with fingernails rather than claws. Forming complex social groups, these early primates began to develop larger brains, and for millions of years, they remained in the trees. Then about 5 million years ago, the climate began to change, and the forests receded as grasslands expanded across eastern Africa. Adapting to the changing landscape, our primate ancestors came out of the trees and onto the grasslands. They began to walk upright, possibly because it made it easier to spot predators, or because walking on two feet was a much more efficient way of moving. Either way, it freed their hands for other purposes, notably making tools.

Eventually, modern humans evolved around 200,000 years ago in Africa; a date that has been supported by both fossil and DNA evidence. Although, this date is being challenged by new fossil evidence, pushing the date of first humans to about 300,000 years ago. Additional fossil and DNA evidence indicates that populations from Eastern Africa migrated and populated the rest of the world. In the last 10,000 years, humans have been incredibly successful as we have developed new technologies aiding in our survival and proliferation in practically every part of the world. In 2018, there are 7.7 billion humans on the planet, making us the most abundant large mammal on the Earth.

Chapter 11

Life Interacts
With the Environment

The day to day ecological interactions between individuals
and their environment is natural selection
driving adaptive evolution.

Introduction

No living thing is an island, isolated from all other life. Every organism must interact with the living (biotic) and the non-living (abiotic) parts of their environments. Ecology is the study of the relationships between living organisms and their environment, and the scope of ecological studies is quite vast. At its smallest scale, the focus may be limited to studying populations of a single species. In contrast, large-scale ecological studies focus on global patterns of diversity and productivity lasting decades.

Ecology, like any branch of science, starts by asking questions. Why do we find different plants and animals as we move across the landscape? Why do pines, spruce trees, and other conifers dominate the western forests of North America while deciduous trees dominate the eastern forests? Why are Greater Roadrunners common in the southwest, but not in the eastern U.S.? To answer these types of questions, ecologists rely on natural history, an understanding of the distribution of plants and animals to study their relationships to each other and their environment. Natural history is mostly an observational science relying less on experiments and hypothesis testing.

Every species resides within a geographical range where you can find it, in some places they are more common than others. Ecologists define a species place in an ecosystem as its niche. A species tolerance to physical conditions in the environment and biological factors, both of which work together to determine where you may find a certain species and define its ecological niche. Additionally, the deep history of the Earth and continental drift also plays a significant role in where we find certain types of animals.

The connection between ecology and evolution is important, the species we see today are the result of selection in the past. It's also important to understand that every adaptation comes at a cost, a trade-off in that you do something else less well. Therefore, there are no perfectly adapted species. That's why you wouldn't find a desert adapted lizard thriving in a rainforest.

The Importance of Natural History
For Understanding Ecology

Ecology begins with an understanding of natural history, a science that relies on observations to learn how organisms in a particular area are influenced by their physical surroundings and other organisms. Birds are among the best-studied organisms in the world as legions of birders contribute millions of observations of bird sightings to a single data-base called *eBird*, which is run by the Cornell Lab of Ornithology.

For example, if you're a birder, then you know to look for Black Oystercatchers on rocky outcroppings along the Pacific coast, or you would search for a Grace's Warbler in the canopy of a Ponderosa pine tree in the Western United States in the late spring through summer. Likewise, you know that you will only find Limpkins, a medium sized brown bird, in Florida where apple snails are abundant and it doesn't get too cold in the winter. Birds also undergo seasonal variation based on their migration patterns. In the winter, ducks and geese migrate southwards to overwinter in the southern states to avoid freezing water. When the temperatures rise in the spring, they return to their summer breeding grounds. Some birds including, Greater Roadrunners, White-winged Doves, Western Scrub Jays, and House Finches are year-round residents in many western states. Sometimes, I go to the beach on the spring high tide to watch thousands of shorebirds congregate at the water's edge to eat the eggs of horseshoe crabs.

From thousands of observations, we build geographical and seasonal range maps of individual species. Range maps for a particular species become more accurate when they include suitable habitat. If you've ever picked up a where-to-go birding guide, it helps you locate hard to find species by sending you to the right habitat at the right time of year. To further improve your chances of finding rare birds, a good field guide may include specific behaviors and habits of the bird that has eluded you. For example, a Brown Creeper is usually found on the trunks of trees working their way up towards the crown. A Spotted Towhee usually forages on the ground in thick underbrush scratching the leaves searching for a meal. Want to find a Vermilion Flycatcher? Watch for reddish birds in the distance sallying for insects in open areas.

These types of natural history observations are not limited to birds, naturalist have been collecting observations on many types of plants and animals. For example, the ground-dwelling whiptail lizards (*Aspidoscelis*) are active on the ground when it is hot. Different species of anoles may be found on grass, on small limbs, on the trunks of larger trees, or high up in the canopy. Plants too have different distributions. The columnar cacti of the Sonora desert aren't found in the drier Chihuahua desert, and they aren't found to the north or in higher elevations where it gets too cold. If you've ever driven into the high country of the Rocky Mountains from the eastern plains, the terrain transitions from grasslands, to juniper and pinyon, to Ponderosa pines that eventually give way to mix coniferous forests dominated by fir and spruce trees. Above 11,000 feet, the trees abruptly end at the tree line where the alpine tundra begins. Similar to eBird, iNaturalist allows you to submit observations and photos of any organism.

The distribution of plants and animals described above is determined by both the physical environment and their interactions with other living organisms. Recall that no organism is an island. Therefore, life interacts with the non-living (abiotic) and living (biotic) part of their world to survive. Over the course of evolutionary history, the number of species has continually increased, thus increasing the types of ecological interactions and the complexity of ecosystems. Where natural history is mostly observational based, ecology relies more on question driven hypothesis testing to better understand how ecosystems are structured.

An important concept in ecology is the niche. The niche of a species is where and what that species is doing in an ecosystem. But, what are the factors that determine a species' niche? We can divide the niche of a species into the fundamental niche and realized niche. In broad terms, the fundamental niche is where a species can be found based on the physical environment or abiotic factors, including climate, disturbance, soil type, salinity, *etc*. The realized niche is where you actually find a species based on biotic interactions, including competition, predation, diseases, symbiosis, *etc*. The realized niche must be the same size or smaller than the fundamental niche for any species.

The Physical Environment
Determines the Fundamental Niche

Abiotic factors and the physiological tolerances of a particular species determines its fundamental niche by limiting where it can live. The abiotic factors affecting the distribution of different species arise from the physical environment. Some important factors include the climate, types and frequency of disturbance, soil types, nutrient availability, amount of light, and salinity (think marine versus freshwater ecosystems). Many of these abiotic factors also vary over time. For example, the climate of temperate regions experiences cold wet winters paired with hot summers. In the southwest where monsoonal rains are important, the first half of the summer is dry and the second half is wet from frequent thunderstorms.

A species' physiological tolerance to its environment places limits on its geographical distribution and the time of year you can find it. Freezing temperatures place a major limitation on the distribution for many animals. For example, ducks can easily withstand freezing temperatures because their feathers insulate them from the cold. However, they must migrate southward in the winter because the ponds where they breed and forage in the summer freeze over in the winter. Tropical and subtropical organisms could grow well during the summer in most northern latitudes; however, freezing temperatures in the winter will kill them. That is why tropical species remain confined to the tropics. If you are a freshwater fish, then you can't swim into the ocean because the higher salinity would cause you to lose water. Likewise, a marine fish swimming up a river would absorb too much water.

Climate is one of the most important abiotic factors determining where a species can live. However, determining the fundamental niche of a particular species can be difficult because several abiotic factors could be important and vary over time. Additionally, biotic interactions, such as competition, can further restrict a species range. Below, I focus on how climate and disturbance determine the fundamental niche of a species.

Climate

Climate is at the heart of determining the geographical range for many species because of physiological tolerances. Climate is the average weather for a region. It includes the average yearly precipitation and average temperatures; along with seasonal variations including winter lows and summer highs, or the presence of dry versus wet months. Weather is basically the high and low temperature plus the precipitation for any given day. Weather is what you get today or tomorrow, or the daily fluctuations of the atmosphere.

An important difference between weather and climate lies in their predictability. It may be impossible to predict the weather more than a few days out. However, making predictions based on climate is much easier. For example, if you live in Florida, then on any given day in July the temperature will most likely be in the middle 90s with a chance of rain. However, you cannot predict with any certainty when and where a particular thunderstorm may form.

To understand how climate determines were a species can live, let's take the Ponderosa Pine, one of the most common trees in the western United States. If you live at the base of the Rocky Mountains, Ponderosa pines are largely absent and grasses and shrubs dominate the ecosystems. However, if you drive or hike into the mountains, you would begin to see large pines once you reached about 7500 feet in elevation, give or take a few hundred feet. At this elevation, the average temperature is lower and the annual precipitation also increases as you go up a mountain. The connection here is that as you go up the mountain, the climate changes and becomes favorable for Ponderosa pines. At lower elevations, the climate is not favorable for Ponderosa pines, it becomes too hot and dry for this tree, so its distribution is limited by climate.

Sometimes the climate for a particular region can be complex and the organisms living there must be adapted to changing conditions. For example, rainfall can vary dramatically throughout the year. Nowhere is this truer than in the desert southwest of Arizona and New Mexico. The first half of the summer is typically hot and dry, whereas the second half receives monsoonal rains starting in early July and ending in September.

In the southwest, plants and animals time their life cycles with the onset monsoonal rains. For example, spadefoot toads and tadpole shrimp develop quickly in temporary pools. The tadpoles of spadefoot toads grow rapidly and emerge as small froglets before the temporary ponds dry up. After the rains, they use specialized spades on their hind legs to burrow deep into the ground where they wait for the next monsoon season. Coinciding with the spadefoot toads are small primitive looking tadpole shrimp. Their life cycle takes about two weeks, it starts when the eggs become immersed in water, then they rapidly grow into adults, mate, and lay eggs all within about two weeks during the monsoon season.

In temperate regions, summers and winters form distinct seasons. Contrary to popular belief, the distance from the sun does not determine summer and winter climates. Instead, the Earth is tilted on its axis changing the angle of incoming light as it orbits the sun. Winter occurs in the northern hemisphere when it is tilted away from the sun and summer occurs when it is tilted towards the sun. When the northern hemisphere is pointed away from the sun, the incoming light arrives at an angle so that less energy hits the Earth. When combined with shorter days, the temperatures drop, reaching their coldest temperatures typically in January. Because of the large temperature differences between summer and winter, organisms living in temperate regions must be adapted to deal with the cold or migrate to warmer regions.

Plants and animals respond in different ways to the seasonal changes in temperature. Deciduous trees lose their leaves, completely shutting down photosynthesis for the winter. Once temperatures warm in the spring, they once again grow new leaves. Losing leaves in the winter provides several benefits including avoiding damage to the leaves if they freeze. Also, in areas with snow, the plants won't accumulate as much snow and ice on their branches that could potentially damage the tree by breaking limbs. Plants lose water through their leaves, if the ground were to freeze, the available water would be near zero, and the plant would face desiccation even with plenty of water in the ground. One last advantage is that plants would face less herbivory in the winter with no leaves.

A disadvantage of being deciduous is that plants lose energy when shedding their leaves or when competing with an evergreen tree when it begins to warm in the spring. Deciduous trees must grow new leaves to begin photosynthesis, whereas an evergreen can immediately begin to grow when conditions are favorable.

Some mammals deal with colder temperatures by hibernating, allowing their body temperature to drop so they don't have to spend as much energy maintaining a high body temperature. Large mammals like Brown Bears feed all summer long, building up fat stores they slowly burn over the long winter. Hibernating mammals use brown fat that specializes in producing body heat by burning up their fat storage, but doesn't produce a lot of ATP.

Although, birds are known for making long migrations, many people remain unaware of how many birds actually migrate in the spring and fall. Birds are among the most active animals on the planet with very high energy demands. To supply their energy, birds require energy-dense foods, that's why most are predators, or seed eaters. Few birds can survive solely on eating leaves. If they do eat plant material, it usually includes energy dense seeds and fruits.

During the winter, insects become quite scarce, especially in areas where it routinely freezes. To follow resources, many birds migrate south, tracking warm weather and abundant food sources. In North America alone, it's been estimated that approximately 5 billion birds migrate to Central and South America in the winter and then back to North America in the spring. One shorebird called the Black-tailed Godwit migrates from Alaska to New Zealand in one trip, traveling 8,000 miles across the Pacific and losing over a third of its body weight along the way!

The location of continents, oceans, mountain ranges, and prevailing wind patterns also play a role in the climate of a region. The atmosphere and the oceans move energy away from the tropics, towards the poles. Large atmospheric circulation patterns known as Hadley cells move air from the tropics to about $30°$ latitude north and south. The air near the equator is warm and humid, as it rises, it cools and drops the moisture as rain. That's why it rains so much in the tropics and forming tropical rain forests confined to regions near the equator. The cool dry air moves away from the tropics and falls around $30°$ north and south latitude preventing precipitation, that's why large deserts are located at similar latitudes around the world.

Large mountains create local and regional variation in climate. When you move up a mountain, the average temperature decreases while precipitation increases. These changes are caused when prevailing wind currents encounter a mountain and the air is forced upward. Similar to the tropics, the air will rise and cool causing more rain or snow. That's why larger mountains often have forested ecosystems with higher elevation. In Albuquerque, the average precipitation is 12 inches per year, but the Sandias, rising over 5,000 feet above the city, may receive 30-40 inches of precipitation per year. Large mountains can remove enough water to create much drier conditions behind them. Nowhere is this more evident than in India where air currents hit the Himalaya Mountains, which forces them to drop their moisture causing monsoonal rains. North of the Himalayas, the climate becomes much drier. In fact, the driest desert in the world is the Atacama Desert in Chile, lies in the rain shadow of the Andes Mountains.

Disturbance

Along with climate, disturbance plays an important role in determining where species can exist. Patterns and types of disturbance can be complicated. In general terms, disturbance is a temporary change to the environment associated with the removal of biomass from the ecosystem. The removal of biomass is the loss of organic matter; if a fire burns the fuel on the ground, then there has been a loss of biomass. Most disturbances including fire, flooding, grazing, or high winds may act quickly lasting a few hours to a few days. Other disturbances, including a prolonged drought, may last weeks to several years. Contrary to popular belief, natural disturbances, including fire and flooding, are important for maintaining high levels of diversity as many plants and animals depend on disturbances. When natural disturbances are removed from ecosystems, it can be followed by a loss of diversity as it transitions to other common species.

Fire is perhaps the most demonized and least understood type of disturbance in the public eye. Decades of Smokey the Bear telling us that "Only you can prevent forest fires", and fires kill Bambi has led to fire suppression across the west. The demonization of fire has been unfortunate because fire is an integral part of many ecosystems, including grasslands and most coniferous forests in the west, far north, and the southeast. In these regions, many plants are well adapted to frequent fire and in many cases, depend on it to complete their life cycle. For example, Sand Pines (*Pinus clausa*) in the southeast open their cones only after being heated by fire, meaning new trees can't grow unless there has been fire. Longleaf pines (*Pinus palustris*) require open areas with lots of sunlight for their seeds to germinate. Fire removes other plants and debris, making it easier for the seeds to germinate. Longleaf pines are so dependent on fire that even their needles have evolved to easily burn, but not too hot so that young plants aren't damaged by the fire.

The loss of fire has had profound effects on our forests. Over a hundred years ago, the southeastern United States from the Carolinas to Texas was dominated by open park-like Longleaf pine forests where ground fires spread through the region. Sometimes the fires would last for weeks burning hundreds of thousands of acres. These fires would occur on a regular frequency of about 2-4 years, removed fuel on the ground, and prevented the succession to a hard wood forest by killing young hard woods. As a result, the forests remained open and park-like with scattered trees and an understory dominated by wiregrass and wildflowers. From early spring to late fall, the understory of Longleaf pine forest would be packed with white, yellow, red, and purple flowers. In some cases, the diversity of plants would reach upwards of 50 species per square meter (a little over 9 square feet).

Beginning in the 1920s, these natural occurrences of fire were suppressed; not just in the southeast, but throughout the western states as well. In the southeast, the loss of fire shifted the ecosystem from open forests dominated by Longleaf pines and wildflowers to thick hardwood forest dominated by various species of oak trees. The transition caused the loss of plant and animal diversity. In the west, fuel built up on the ground and the density of trees increased leading to more intense catastrophic forest fires that kill all the trees, requiring decades for the forest to return.

The U.S. Forest Service along with other government agencies have been working hard in the last few decades to restore fire as a disturbance to ecosystems by conducting controlled burns to remove fuel and improve diversity. But these efforts are being slowed by rampant urban expansion fueled by population growth, letting many people live near fire-dependent forests. Unfortunately, people living near national forests often don't like the smoke from the controlled burns, making it difficult to properly manage our forests and restore the ecosystems to a natural state. The irony is that the lack of controlled fires increases the chances of property damage from fire.

Flooding: Similar to fire, river flooding is usually considered something bad because of the property damage that results from flooding. To control flooding, various government agencies built nearly two million water control structures in the U.S. alone. But, like fire, flooding is a periodic disturbance and a natural part of many aquatic systems and their surrounding wetlands. Organisms in the streams and adjacent wetlands depend on regular flooding for their survival. In the Rocky Mountains, spring flooding occurs when the snow pack melts causing streams and rivers to swell and flood their banks. Importantly, the high water replenishes underground water supplies and it also removes biomass and creating bare areas, which allows plants to seed and grow. The routine flooding prevents any one species from taking over and dominating the landscape, thus preserving diversity.

In New Mexico, along the banks of the Rio Grande River, is a ribbon of forest dominated by cottonwoods (*Populus fremonti*) and called the Bosque (Spanish for forest) by the locals. Historically, the Bosque was flooded each year by spring runoff from snow melt at the headwaters located in the San Juan Mountains of southern Colorado. The spring flooding removed debris, creating bare ground for the cottonwood seeds to land and germinate. Starting the in the 1940s, water control devices including jetty-jacks and dams were put in place to prevent the natural overbank flooding. As a result, cottonwood seeds cannot germinate to replace older trees as they die. Over time, the large cottonwood trees will be replaced by salt cedar and other non-native trees, further reducing diversity in the Bosque.

Other Abiotic Factors
Light, Nutrients, Salinity, and soil Type

The amount of sunlight and nutrients available determines the rate of photosynthesis. This becomes especially important in the oceans where light doesn't penetrate much beyond 700 feet, which is less than 10% of the ocean's average depth. The vast majority of our ocean's environment remains in perpetual darkness. Ironically, much of the ocean's surface is a biological desert despite having abundant sunshine and water. Being far from land, these regions lack vital nutrients required for phytoplankton growth. In this case, nutrients limit where the species are found, not water or energy.

The salinity of a system also places limits on where aquatic organisms can live. In fact, the entire phylum Echinodermata is completely limited to marine environments. Not a single species of starfish, sea urchin, or sea cucumber has ever successfully evolved to survive in freshwater. There are more than 30,000 species of fish, approximately half are confined to freshwater environments while the other half are marine. The differences in salinity has resulted in few fish capable of surviving in both marine and freshwater environments.

On land, soil types play an important role in determining where plants can grow. Plants that live in sandy, well drained soils must be adapted to handle periodic dry conditions. Another factor, the pH of soil effects nutrient availability to plants. Wetland areas often have acidic soils lacking nutrients, or at least they are not available to plants. These areas have carnivorous plants, that supplement their nutritional needs by capturing small animals.

Biological Interactions Determine the Realized Niche

The realized niche of a species is where you actually can find it. A species' realized niche will always be a subset of the fundamental niche. Many types of biotic interactions including herbivory, predation, mutualism, and parasitism determine the realized niche of an organism. Biotic interactions can be harmful, neutral, or beneficial, which ecologists describe this as -/-, +/-, +/0, or +/+ relationships. Below, I discuss how biotic interactions determine the realized niche for different species.

Competition (-/-)

I'll never forget exploring a rock outcropping along a remote beach in the Bahamas, it was covered with different species of colorful snails of many sizes. The smaller ones wedged themselves in tiny cracks while the larger ones were on the upper surfaces unable to get into the smaller cracks. Regardless of their sizes, these snails were all grazers, scraping algae off the rocks. Because of their size differences, they coexisted on the same rocks by exploiting different microhabitats. In this example, the larger snail, a Nerite, has a realized niche similar to its fundamental niche. The smaller snails were confined to small crevices because they could not directly compete against the much larger nerite snails. As a result, the larger snail limited the realized niche of the smaller snail species due to direct competition. By evolving into different size classes, these snails lived on the same rock by reducing competition between the species, they just utilized different microhabitats.

Similar types of competition are common throughout the world. If you visit a beach, watch for sandpipers foraging along the shorelines. They stick their bill into the sand probing for small invertebrates. On a single beach, you may find a dozen or more species, each with different body sizes, bill lengths, and habits. Over time, natural selection has driven the evolution of sandpipers into unique species to exploit different resources on the same beach to reduce competition between species, also known as interspecific competition.

Sanderlings and Marbled Godwits provide a good example of how two species live on the same beach by dividing the resources, thus occupying different niches. The Sanderling is a small white bird with a short straight bill often seen running just in front of the waves crashing on the beach. They go after animals living near the surface of the sand, whereas the much larger Marbled Godwit will use their long bill to probe much deeper into the sand, finding different prey. These species divide the resources of their environment by exploiting different prey items based on bill size allowing these birds to live on the same beach. Other sandpipers may forage further away from the surf, while others may prefer mudflats in bays and inlets.

Competition is not limited to animals, plants also experience lots of competition. Typically, you will find Ponderosa Pines at mid-elevation in the Rocky Mountains. However, they can grow at higher elevations. The trade-off is slower growth and its larger branches become easily damaged in years with lots of snow. In this case, the narrowly-branched fir trees out-compete the pines because their branches accumulate less snow, thus avoiding damage from excess snow. In this case, the Ponderosa Pine, can live at higher elevations, it's just not found there because it is out-competed by fir trees which are better adapted to the colder snowier conditions.

These examples of competition between species shows how competition restricts where a species is actually found. Direct competition between species, costs both species and is therefore considered a -/- interaction. Competition reduces the size of a species realized niche. It also shows how day-to-day interactions and natural selection drive the evolution of species to reduce direct competition. In the case of the snails and sandpipers, competition between species drove the evolution of two different shell sizes, or different sizes of bills. This makes sense because direct competition between species is negative for each species. Although similar species may live in the same areas, they are utilizing the habitat differently based on different traits that were the result of natural selection driving their evolution to reduce competition. As a result, they have different realized niches.

The evolution to reduce competition always involves trade-offs for each species. The larger snails can eat larger pieces of algae, displacing the smaller species, but their large size prevents them from exploiting the resources found in tiny rock crevasses. Fir trees have smaller branches that don't break during heavy snow falls. However, at lower elevations they are out-competed by the Ponderosa pines with larger branches allowing the pines to gain more access to sunlight. As you can see, evolution by natural selection is about trade-offs, if you evolve to do one thing well (reduce snow accumulation on your branches), you become less well adapted to do something else (growing larger branches to capture more sunlight). Therefore, there are no perfectly adapted species.

Predation (+/-)

The rise of predators at the start of the Cambrian was a major factor behind the evolution of animals. Even today, predation remains a driving force of evolution. Predation occurs when an animal (or even plant) kills another animal for food. Predation directly benefits the predator (+) at the costs of the prey (-), hence the +/- relationship.

We are most familiar with large charismatic predators like Cheetahs, Lions, Orcas, and Great White Sharks, but the world is full of predators in all shapes and sizes. They employ many tactics to catch prey; Cheetahs, Peregrine Falcons, and Blue Marlin use speed to catch their prey. Spiders build intricate webs to catch flying insects. Vipers and cobras inject specialized venom to kill their prey, while boas and other constrictors suffocate their prey. South American leaf fish look just like a leaf, so much so that their prey doesn't recognize the danger as the leaf fish slowly moves in. Once its close enough, it rapidly opens its mouth sucking in the unsuspecting prey. Even more strange are the electric eels and catfish that generate a powerful electrical field so strong they stun their prey to capture it without a fight.

Perhaps the strangest looking predators are the deep-sea anglerfishes that spend their lives in near to total darkness. The most prominent feature of these predators are their enormous mouths full of sharp teeth, an image straight out of a science fiction horror story. They also use modified a dorsal fin that houses bioluminescent bacteria to attract prey. In the depths of the oceans, an encounter with a potential prey is rare, so these fish have large mouths with long teeth, so they can grab rather large prey. This way their meal will last them a long time, or at least until their next encounter.

The mode of predation limits a species habitat preference. For example, Peregrine Falcons are known for their incredible speeds in the air as they dive-bomb medium sized birds such as ducks. However, they lack maneuverability, preventing them from catching small woodland birds in thick forests. Likewise, predation limits the range of prey. If they get pushed to the margins of their habitats, predators easily pick them off. Small woodland birds may be very agile in flight, but if they flew into the open, they would be relatively easy targets for fast moving falcons.

Symbiosis

Symbiosis means living together, a symbiotic relationship occurs when two or more different species have close interactions, and often for long periods of time. The Disney film *Finding Nemo* popularized the symbiotic relationship of clownfishes with their anemones. Symbiotic relationships can be mutually beneficial when both species benefit (+/+), commensal when one benefits while the other is unaffected (+/0) or parasitic when one species benefits at the cost of the other (+/-). In some cases, the symbiotic relationships are obligatory, meaning that both species must be present for either one to survive.

Mutualism: Ants are an amazing insect; found on every continent except Antarctica, they are vital to the functioning of ecosystems. In the tropics of the Central and South America, leaf cutter ants harvest leaves from the surrounding forest to grow a special fungus that they harvest in their underground gardens. The fungus benefits from the constant supply of food brought in by the ants, and the ants benefit from the nourishment they receive from the fungus.

Leaf cutter ants are not the only ants to form mutualistic symbiotic relationships. If you have a flower garden, look closely and you may see small aphids and ants on the stems of your plants. After a few careful observations, you will see the ants tending to the aphids and protecting them from predators. Because the aphids insert their rostrum, a specialized mouth part, into the plant's vascular tissue, they constantly secrete a sugary liquid eaten by the ants. In this case, both species benefit from the relationship.

One of my favorite mutualistic relationships occurs in corals, animals related to jellyfish and sea anemones. Coral reefs are home to some of the most productive and diverse areas on the planet. In fact, coral reefs may account for nearly half of all marine species, yet they occur in less than one percent of the oceans. Surprisingly, coral reefs are only found in crystal clear waters devoid of many nutrients resulting in a paradox of how you can have so much productivity and diversity in very nutrient poor areas. Recall from Chapter 9 that the answer lies in the symbiotic relationship between the coral, a small animal, and a single-celled photosynthetic organism called zooxanthellae that lives inside the coral. The coral sends out its tentacles to catch small prey or debris in the water supplying vital nutrients, while the symbiotic zooxanthellae produces carbohydrates through photosynthesis. Without this mutually beneficial relationship, tropical reefs would not exist and there would be much less diversity in the world.

Parasitism: Parasites form symbiotic relationships with their host, or in some cases multiple hosts to complete their life cycle. Some parasites manipulate their host, altering their behavior to ensure the continuation of their life cycle. The liver fluke called *Dicrocelium dendriticum,* which infects cows, provides a striking example of complex life cycles host manipulation. Inside the liver, the worm mates and lays eggs that are excreted in the cow's feces. Small snails eat the eggs that hatch and reproduce asexually inside the snail. However, the snail will get rid of the parasite by excreting them in a ball of mucus. At this point, ants eat the snail's mucus and ingest the flukes that will then move to the head of the ant. Once there, they alter the ant's behavior making them crawl up a blade of grass in the evening to get eaten by a cow, repeating their life cycle. If the ant doesn't get eaten, the fluke will let the ant carry on with its normal business until the next evening.

History Matters

Have you wondered why Australia is full of unusual marsupials, such as koala bears and kangaroos? It turns out that when it comes to the distribution of animals such as marsupials, geological history matters. The dominance of marsupials in Australia is not so much of what happened a few hundred years ago, but what happened hundreds of millions of years ago and continues today. Continental drift, the slow movement of continents across the planet over tens of millions of years, explains Australia's marsupials.

About 250 million years ago, all the continents were merged into one giant super-continent called Pangaea. Slowly, over millions of years, the Pangaea broke into smaller continents that we know today, including Antarctica, South America, North America, Africa, Europe, Asia, and Australia. Mammals first appeared nearly 200 million years ago during the Mesozoic Era, the earliest ones laid eggs similar to reptiles. Later, the marsupials evolved live birth and spread across the continents. Marsupials migrated from South America to Antarctica and then to Australia. When Antarctica and Australia broke away from the other continents it isolated the ancient egg laying mammals and marsupials from modern mammals. As a result, the marsupials diversified into unique animals including kangaroos and koala bears.

Too big to be an island, too small to be a continent, New Zealand is a continental fragment. It broke away from Australia and Pangaea before mammals had evolved and has existed for over two hundred million years without mammals, except for bats that arrived by flying and some invasive rodents brought by man. Similar to Australia, New Zealand is home to evolutionary relics like the Kiwi and a unique lizard-like reptile called the Tuatara. It superficially resembles a lizard, but has been separated from lizards for over 220 million years ago. Tuataras are a living fossil, replaced on other continents by modern lizards. The long separation of New Zealand from other continents has allowed this ancient reptile lineage to exist for over 200 million years.

Madagascar is another continental fragment that broke away from Africa nearly 60 million years ago, carrying with it early primates. Humans along with monkeys and great apes form a group of primates called the Simians. However, on Madagascar, approximately 100 species of lemurs, an ancient lineage of primates called Prosimians still survives today because they remained isolated from other modern primates. Unfortunately, many are highly threatened or endangered due to habitat loss and hunting pressures.

For the past 2.5 million years the Earth has been in an ice age, a period of long-term cooler temperatures, continental ice sheets, and permanent ice at the poles. Although some geologists argue we have been in an ice age for nearly 35 million years. In fact, we are still in an ice age, just not a period of maximum glaciation. During the last glacial maximum that ended a mere 11,700 years ago, large glaciers covered vast areas North America and Europe. The ice sheets locked up so much water that sea levels were approximately 300 feet lower than today. In North America, glaciers covered much of Canada and extended southward to Kentucky. When they receded, plants and animals colonized the bare lands left behind. Small isolated pockets in the Appalachian Mountains that remained glacier free formed refuges for many species that spread after the retreat of the glaciers.

During the last glaciation, lower sea levels connected many islands currently separated by water today. In the 1850s, a British Naturalist named Alfred Russel Wallace traveled throughout the Indo-Pacific collecting animals and plants for museum collections. He noticed that Australia, Tasmania, and New Guinea shared similar species, but there was a big change in species composition between New Guinea and Borneo. He also made similar observations on several islands where the plants and animals were more similar to Southeast Asia. Now we know that the difference in species resulted from historical connections between New Guinea and Australia during the last ice age, or between islands and Southeast Asia. We call the line between the two regions the Wallace line. During the last ice age, the two regions remained separated by a deep-water channel preventing the mixing of Australian and Southeast Asian species.

What Limits Population Growth?

For the first part of this chapter, I focused on the factors that affect the distribution of species, but what factors limits the growth of populations? The easy answer to this question is that birth rates and immigration will add to a population while death rates and emigration will reduce population size.

The more complex answer lies in what determines death and birth rates or emigration and immigration rates. Over two hundred years ago, the English scholar Robert Malthus understood that any population can grow exponentially. It goes something like this, imagine a single *E. coli* bacterium that divides into two daughter cells, then each daughter divides and there are four cells, then eight, then 16, 32, 64, 128, 256, 512, 1000, 2000, 4000, 8000, 16,000, 32,000 and so on. Within a few days, the Earth would be covered in *E. coli*.

Exponential population growth is not limited to rapidly dividing bacteria. If you were to take a single pair of houseflies at the start of summer, if all their offspring and all their offspring's offspring (grandflies) were to survive throughout the summer, there would be more than 325 trillion individuals, or enough houseflies to circle the Earth's equator 57,000 times! Luckily for us, we are not covered in houseflies or *E. coli*. What determines birth and death rate, or immigration and emigration?

Ecosystems have a maximum amount of resources available, there is only so much energy, nutrients, and space available. Taken together, the available resources determine an ecosystem's carrying capacity, which is the maximum population size that can be reached. In many areas, carrying capacity fluctuates based on climate patterns in the area.

Populations of plants and animals will respond to changes in the carrying capacity as environmental conditions fluctuate. When resources become abundant, the carrying capacity of an ecosystem increases and populations will grow. Sometimes, a population can grow too quickly, overshooting the carrying capacity of the ecosystem. When this happens, the population will decline. Likewise, if resources become scarce, as happens during drought, then populations respond by becoming smaller as individuals either die or immigrate to other areas.

Deserts provide a good example showing how carrying capacity fluctuates annually. Deserts receive plenty of energy from sunlight; however, water limits their carrying capacity. In the Southwestern U.S., deserts depend on monsoonal rains beginning in July. In wet years, the rains lead to more plant growth, which in turn provides more food for herbivores of all types from insects to antelopes. When populations of insects go up, the populations of insect eating birds respond by growing too. Therefore, in wet years, the carrying capacity for the desert becomes higher with more water. In dry years, plants grow much less producing fewer seeds and fruits. This makes for fewer herbivorous insects leading to fewer insect eating birds. For example, in parts of New Mexico between 2010 and 2014, Burrowing Owl populations drastically declined, a direct result of prolonged drought in the region reducing the area's carrying capacity for this species. The drought drastically reduced plant growth, which in turn caused a decline in rodent populations the main food source for Burrowing Owls.

The ability of a species to respond to changes in the carrying capacity varies based on the length of their life cycle, and number of offspring produced during a given time. Insect populations may fluctuate wildly from year to year due to their rapid generation times. Black Bear, on the other hand, have much longer life spans and lower reproductive rates responding more slowly to changes in carrying capacity. If their populations were to become severely reduced, it may take them several years to recover.

How quickly a species responds to fluctuations in carrying capacity depends in large part on its survivorship curve. A survivorship curve predicts the proportion of individuals surviving to a certain age class. In general, there are three different survivorship curves, Type 1, Type 2, and Type 3. Humans and whales exhibit a Type 1 curve; we have a high probability of making it from one year to the next. Over 50% of people born will make it to the average maximum age of the species. As individuals begin to approach the maximum age, the rate of mortality from one year to the next drastically increases. In a Type II curve, individuals have about the same chance of surviving from one year to the next. For example, a squirrel may live a maximum of about 5 years, but each year, it has a 20% probability of death. Animals with a Type III curve experience the greatest mortality early in life. Spadefoot toads lay hundreds of eggs, but most tadpoles and young frogs die in the first year. However, once they make it to adulthood, they will live a long time.

The Day to Day Ecological Interactions
Drive Adaptive Evolution

Evolution, the change in a species over time, is a fact because it actually happens. We can observe it, measure it, and experimentally verify that it is taking place. Darwin's theory of evolution by natural selection provides the best explanation why species change over time, or evolve. Recall that natural selection makes several assumptions, the first that there is variation within a population where some individuals are more fit to the environment. Second, life and death are not completely random, at least not on average, therefore some individuals are more likely to survive, reproduce, and pass on their favorable genes to the next generation. Over time, favorable traits accumulate in the population causing the species to evolve.

Darwin originally assumed evolution was a slow process. However, the rate of evolution can be quite rapid like the evolution of antibiotic resistant bacteria, or imperceptibly slow as seen with the horseshoe crabs, whose basic form has not changed in nearly 450 million years! Although modern horseshoe crabs look like their ancient ancestors, they are still undergoing evolution. It's much a like a Volkswagen car from the 2000s looks superficially similar to one built in the 1960s, but basically every aspect of the electronics and engines have changed to become more efficient.

What does it mean to be better fit to the environment? It comes down to the day to day survival of finding food, avoiding predation, fighting off diseases and parasites, and reproducing. If the climate you find yourself in becomes too hot, do you have ways to remain cool, like sweating or panting? Can you search for shade or bury yourself underground? If it becomes dry, can you conserve water, or easily find new water sources? If predators are present, can you outrun them, avoid being seen, or fight them off? If you are a group of male prairie chickens on a lek, is your song and dance enticing enough for the females? Are your feathers bright enough to attract the females? All these factors are important for survival and reproduction.

By now, we start to see the day to day ecological interactions place selective pressures on individual organisms. Surviving a cold spell, evading a predator, finding food, and reproducing occur on the short time scales of hours, days, or even weeks. These selective pressures act on the individual, either you survive and reproduce, or you don't. An important point to remember, natural selection works on the individual, but an individual does not evolve, only populations evolve. Sometimes, populations cannot adapt quickly enough to rapidly changing environments and they perish. If all the populations of a species are lost, then it is extinct. It has been estimated that 99% of all species that have ever lived on the Earth have gone extinct. That means that the roughly two million species alive today represent only 1% of all the life that has existed on the Earth!

Every adaptation in a species comes at a cost, that cost is the loss of being able to do something else well. For example, a cactus can survive in extremely dry environments due to its ability to conserve water. It has shallow roots to quickly absorb water from infrequent rains and their leaves have been reduced to thorns serving a two-fold purpose of protecting the plant from herbivores and to prevent water loss. Cactus further reduce their water loss by opening their stomata only at night or early in the morning when the humidity is highest for the day. Stomata are tiny openings that allow gas exchange with the atmosphere and help control water loss in plants.

These adaptations help cactus survive in arid environments where few plants can, but the trade-off is that they grow very slowly. In fact, the saguaro cactus of the Sonora desert may live for hundreds of years. Even if you were to provide it with plenty of water, it still wouldn't grow much faster. Ironically, if you water a cactus too much, it will actually die because it cannot tolerate too much water. Using an animal example, Cheetahs may be the fastest land animal, but that limits them to smaller prey compared to a lion, which weighs almost 4 times more than a cheetah. Cheetahs would not survive well in a forest because trees would prevent them from moving at full speed, allowing their prey to more easily get away. You can't be good at everything at once.

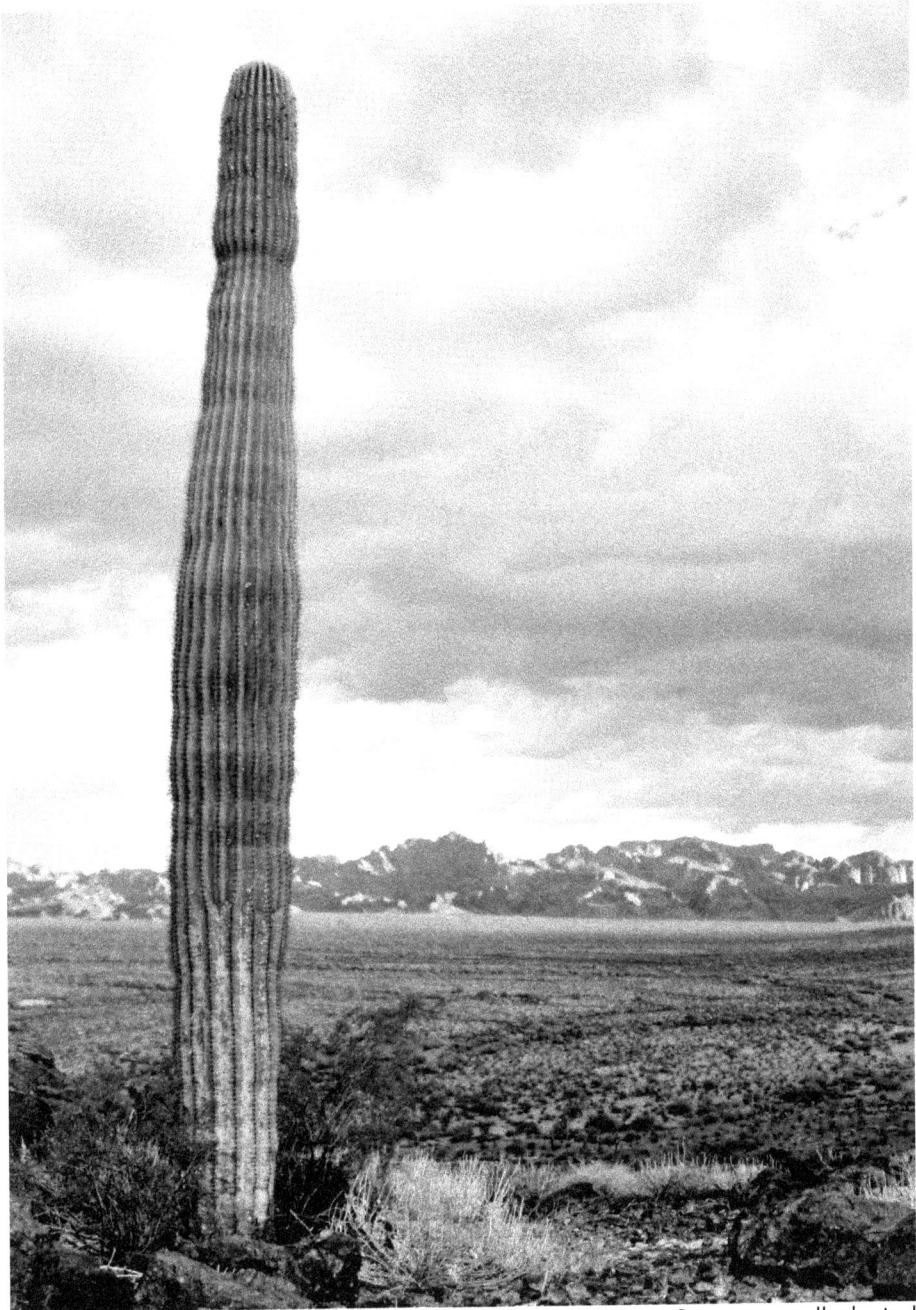

A Saguaro Cactus in the Kofa National Wildlife Refuge in Arizona. Cactus are well adapted to living in desert environments by preventing water loss. However, the adaptations of cacti in deserts is a trade-off, if you supply them with lots of water, they will not grow as fast as other plants adapted to wetter environments.

Chapter 12

Life is Connected to the Earth

When you get right down to it,
Life is an extension of geological cycles

John Rogers

Introduction

Energy flows through ecosystems, making life possible. But, life is more than energy, and contrary to science fiction, you won't find beings of energy. There's more to life. Life uses energy to create order out of chaos by taking elements and combining them to form molecules creating complex systems out of equilibrium with its environment. All the elements and molecules used by life come from the environment and cycle through living organisms making life an extension of geology.

Every living organism must get nutrients and energy from the environment, you must have both. Limit one or the other and life struggles. Ecosystems connect life to the Earth. Energy makes nutrients available to life, think about it this way, every carbon atom in your body came from a molecule of carbon dioxide that was fixed in to an organic molecule by photosynthesis in some plant. As you can imagine, this makes photosynthesis, which relies on energy and nutrients, vitally important to almost every ecosystem on the planet.

All ecosystems must also   recycle materials, including carbon, nitrogen, phosphorous, and water. It's these nutrient cycles that connect life to the Earth as nutrients move back and forth between the environment and living organisms. Most ecosystems rely on plants and other photosynthetic organisms to harness the energy in sunlight to make these materials available to life. As the energy is transferred through different trophic levels in an ecosystem, the energy that entered as sunlight will eventually exit the ecosystem as heat.

Not only is life connected to the Earth through nutrient cycling made possible by a steady input of energy, these same movements of resources connect vastly different ecosystems to each other. Additionally, unique plants and animals live in the transition zones between two ecosystems, adding to the diversity of the landscape.

Similar environmental conditions, such as a wet and warm climate, leads to similar ecosystems even though the species comprising them can be quite different. Similarities in ecosystems let us categorize them into the biomes. You may already be familiar with biomes including deserts, coniferous forests, temperate broadleaf forests, or grasslands.

Living Organisms Form Communities

If you have taken a hike on a trail whether in the mountains surrounded by tall evergreen trees, or in the desert with hardy cacti, then you hiked in a biological community. All the species living and interacting with each other in a certain area forms a community. A community can be small, such as the microbes living inside your gut. Or, a community could be much larger, such as all the fish and invertebrates living in a nearby lake.

We use different methods to define biological communities including the types of species present and their interactions with each other through competition, predation, parasitism, and other symbiotic relationships. In the last chapter we saw how these interactions were influential in determining the realized niche of a species, or where you actually find a particular species. But, all these species must also live in their physical environment, which determines their fundamental niche, or where they can actually be found based on physiological limitations. Therefore, an ecosystem includes the living and non-living components in an area interacting with each other.

When traveling from one region to another, whether driving up a mountain, or driving to the coast, I can't help but notice that diversity varies across the landscape. The types of species change based in large part on the climate and other aspects of the environment. At the base the Rocky Mountains you will find mostly pinyon pines and juniper, while at higher elevations, the forest becomes dominated by slender fir trees adapted to the colder climates. In marine environments off the west coast of North America, you may swim from areas with hard rocky bottoms dominated by sea urchins grazing on encrusting algae and then abruptly enter into a kelp forest populated with thousands of animals. Whether in a marine or terrestrial community, they vary in the number of species, the relative abundance of the species, and the types of species. Communities are dynamic systems, changing over time as some species come and go, or change in their relative abundances.

One common way to describe communities is the species richness, the number of species present. Diversity varies immensely between communities. For example, a tropical rainforest may have 300 species of trees per acre, whereas a northern boreal forest may have less than

10 species of tree per acre. Differences in diversity, such as a tropical rainforest or a boreal forest, are largely due to the climate and other aspects of the physical environment. In the next sections I will explain why rain forests are more diverse than a temperate forest, or a desert.

The relative abundances of species within a community also varies. There may be upwards of 300 species of trees in an acre of tropical rainforest. In these tropical communities, each species occurs in relatively equal proportions, meaning you would find one or two individuals of each species. Whereas, in Longleaf pine forests, the pines have a high relative abundance where the Longleaf pine accounts for 90% or more of the trees in an acre. Temperate deciduous forests have higher tree diversity than Longleaf Pine forests, but 50% of the trees in an area could be comprised of just one or two dominant species.

As an avid birder and general naturalist, I learned a long time ago that most species are rare. Sometimes, I've found 100 species of birds in a day, but over 50% of the birds on my list included just one or two individuals. At my house, I have seen 40 species, but 22 of those, I've only seen once or twice. The answer to why a few species are common while most are rare is complicated. Predators are rare because only about 10% of the energy is transferred from one trophic level to another. Many species specialize on a certain food source, forage in a specific habitat, nest at certain times, all these factors contribute to their rarity

Sometimes, communities depend on one or several keystone species, which are species with a large impact on a community. Kelp forest provide a great example of how keystone species work. Kelp provides structural habitat for about 600 species of fish and invertebrates. This entire community depends on the presence of just one species, the Sea Otter, a keystone species. A keystone species does not have to be the most abundant animal in the community, but they do have a large impact relative to their population. Sea otters are not the most abundant species inhabiting the kelp forest, but they have an enormous impact on it. The story of Sea Otters goes something like this: Sea otters eat sea urchins, a voracious grazer of algae, including kelp. Remove sea otters and the urchins, released from predation will proliferate and eat the kelp at its base. Eventually, the urchins will cause the complete loss of the kelp forest and the hundreds of species dependent on it.

There are other examples of keystone species. Beavers build dams on streams creating natural wetland areas, which provide habitat for species of plants and animals. The Red-cockaded Woodpecker found in mature Longleaf pine forests drills its nesting cavities into live pine trees. The abandoned cavities are used by Eastern Bluebirds, Carolina Chickadees, Brown-headed Nuthatches, flycatchers, other woodpeckers, Eastern Screech-Owls, and even Flying Squirrels. Without the Red-cockaded Woodpecker to make cavities, these species would decline in population from lack of suitable nesting habitats.

All Ecosystems Cycle Nutrients

Life is connected to the Earth as nutrients continually cycle between us and the environment. It's a process that has been ongoing for billions of years starting since life began. The first living organisms were most likely autotrophs using naturally occurring sources of energy to fix inorganic carbon into organic forms. Once life began, nutrient cycling was forever changed. Carbon, for example, could move quickly between the atmosphere and living organisms, or it could be sequestered as rock for millions of years as part of longer geological cycles.

All life requires nutrients, they can be either organic or inorganic. Organic nutrients include organic molecules such as carbohydrates, proteins, lipids, and vitamins. Organic molecules serve many uses, including sources of energy, building blocks to make other organic molecules required for our functioning, or they can speed up chemical reactions. Most organic molecules are eventually broken down to carbon dioxide and water. Inorganic nutrients include small molecules such as water, carbon dioxide, nitrates, ammonia, phosphates, along with various minerals and electrolytes including iron, calcium, potassium, chlorine, sodium, and *etc.* Carbon dioxide is considered an inorganic molecule because it lacks hydrogen atoms. As an animal, we obtain all our nutrients from the foods we eat and the water we drink.

The elements in nutrients must be continuously recycled because they are not being destroyed or created. These nutrients go back and forth between an inorganic reservoir and living organisms. For example, the atmosphere serves as reservoir for both carbon and nitrogen, and the oceans are the major reservoir for water. While there are many types of nutrient cycles, I'm only going to focus on carbon and water.

I'll start with carbon to show how nutrient cycles connect us to the Earth. You have more than a trillion, trillion carbon atoms in your body, each one is more than 5 billion years old. Additionally, every carbon atom in your body was once in the atmosphere as a molecule of carbon dioxide, perhaps as recently as a few weeks ago. Some of those carbon atoms may have once been in a rock 50 miles beneath the surface in the Earth's mantle, or in a piece of coal, or once part of a protein in a dinosaur that lived a hundred million years ago. When the dinosaur died, its carbon was released into the atmosphere as carbon dioxide. Once in the atmosphere, that carbon atom could have been fixed into a carbohydrate through photosynthesis. Later, an herbivore would come along eating the plant and acquiring that carbon atom, only to breathe it out again. And so, the carbon cycle continues, moving between the living and the non-living world as it is continually recycled.

A carbon atom may be in a molecule of calcium carbonate in a rock, carbon dioxide in the atmosphere, or part of a complex organic molecule inside of a cell. Aerobic respiration and photosynthesis constantly cycle carbon between inorganic carbon dioxide and organic carbon. The last type of chemical reaction will determine what type of molecule carbon may be a part of. No matter how many times carbon is cycled through the environment, or how many chemical reactions it has been subjected to, it is still carbon. That's why you have carbon atoms in you that were once in a dinosaur or a rock.

The element carbon is not only important for organic molecules. As carbon dioxide it is an important greenhouse gas because traps heat like a blanket. Therefore, high CO_2 levels cause warmer climates. Life has played a crucial role in moderating atmospheric carbon dioxide levels by balancing its removal with inputs from natural sources, such as volcanism. Carbon dioxide is also removed from the atmosphere when it dissolves into the oceans forming bicarbonate. Small marine organisms use it to form shells of calcium carbonate, a type of rock used for protection. When these tiny organisms die, their carbon atoms exit the quick back and forth movement between living organisms and the atmosphere. Instead, their carbon enters long-term geological cycling taking place over millions of years as it forms carbonate rock known as limestone. Eventually, those rocks may be buried or even subducted into the interior of the Earth where the immense heat and pressure will form carbon dioxide released in volcanic eruptions.

In the last 400 million years, life has removed large amounts of carbon from the atmosphere by forming coal and oil deposits. In fact, our rapid burning of fossil fuels is releasing carbon into the atmosphere that has been out of circulation for hundreds of millions of years. Once released back into the atmosphere, it is once again used by life when it gets fixed by plant into carbohydrates and other organic molecules. And so, a carbon atom gets cycled over and over again, continuously moving between plants, animals, rocks, and the atmosphere.

Another resource that is continually recycled, water is vital for life because it provides the medium for all life's chemical reactions. Unlike carbon, water is a molecule that can be made and broken down through chemical reactions. In your body, you have water molecules potentially billions of years old, perhaps some that were brought to the Earth from a comet before life emerged. In contrast, some water molecules in your body were made just a few seconds ago.

Generally, when ecologists discuss water cycles, they aren't too concerned about water being broken down or created through chemical reactions (remember, chemical reactions do not change the hydrogen and oxygen atoms that form water). Our oceans form an incredibly large and permanent reservoir of water. Over time, water moves from the oceans to the atmosphere, to the land, through organisms, eventually making its way back to the ocean. The vast majority of water molecules will repeatedly make this journey, sometimes lasting only a few days. Other times, water may seep deep into the ground taking thousands, or even millions of years to return to the oceans. When you drink a glass of water, think of the journey those water molecules have been on. If your water came from ground water, they could have been there for decades to millions of years depending on the aquifer. If your water came from a river, then some of those water molecules were in the ocean only a few days ago.

One of the interesting implications of nutrient cycling is that the elements and molecules in our body continuously move between living organisms and the environment. As I discussed earlier, the carbon atoms in your body could have once been in another living organism, or they could have just as easily been part of a rock, or even pumped from the ground and used as gasoline in someone's car. Pick up a rock, the elements that form the rock are the same age as the elements in your body, they have been cycled through geological processes just like the carbon, nitrogen, and oxygen atoms in your body. It's this movement of elements between life and the Earth that makes life an extension of geological processes that have been taking place for billions of years.

Energy Flows through Ecosystems

In *The Empire Strikes Back*, the ancient Jedi master Yoda explains to Luke that the Force flows through life. If he were talking about energy, he would be right. The constant flow of energy powers ecosystems and life. The ultimate source of energy for almost every ecosystem is sunlight, and most of the energy that enters an ecosystem will eventually exit as heat. Energy is unlike nutrients; it cannot be recycled. Recall that annoying second law of thermodynamics, it tells us that every time energy is used, some of it is converted to heat, a relatively unusable form of energy.

Autotrophs form the base of every ecosystem by using energy to make inorganic carbon and other inorganic nutrients like available to the rest of the community. They form the link between geological or atmospheric cycles and living organisms as they bring carbon and minerals into food webs by incorporating them into organic molecules. The first autotrophs were likely primitive prokaryotes, adding hydrogen to carbon dioxide by using natural energy sources in ancient alkaline vents. Even today, autotrophs in thermal vents do not rely sunlight, they extract energy from rocks, minerals, or hydrogen gas. However, all that changed sometime about 3.5 billion years ago when ancient cyanobacteria began using energy in sunlight to fix carbon dioxide into organic molecules with the benefit of storing energy for later use.

Heterotrophs depend on autotrophs for all their energy and nutrients, although, they are quite diverse. In broad terms, we label heterotrophs as decomposers, or primary, secondary, or tertiary consumers based on their position in the food web. Primary consumers include herbivores, eating only plant material. Carnivores would be secondary or tertiary consumers eating herbivores or even other carnivores. Although carnivores don't eat plants, their energy and nutrients still came from plants. Humans are omnivores, animals that eat both plant and animal material.

On average, when one organism eats another organism, whether it's a plant or animal, only about 10% of the energy is transferred, the rest is lost to the environment as heat. Although, the amount of energy transferred does vary depending on the organism's metabolism. For example, endothermic animals, such as birds and mammals, spend large amounts of energy generating heat to remain constantly active. Therefore, only about 1-3% of the energy they acquire is used for growth or stored, that's why we have to eat so much. Ectotherms, such as a snake, may have a metabolic rate 100 times less than a human when corrected for size differences. That's why they can survive for a month on a single meal, whereas we would starve in a few days.

Another way to think about the loss of energy from plants to herbivores to carnivores is to imagine a patch of grass with about 10,000 Kcal of energy, only about 1,000 Kcal of energy could be transferred to herbivores, and 100 kcal of energy could be transferred to carnivores. Based on this energy transfer, it becomes clear why the world is green and herbivores are rare, and predators are even more scarce. Whales, the largest animals on the Earth maintain their enormous size by going to the base of the food chain to feed mostly on krill, a small crustacean that feeds on phytoplankton.

Decomposers break down organic matter, releasing carbon dioxide and inorganic nutrients, such as nitrogen, phosphorous, calcium and potassium, making them once again available to plants. While fungi are the most well-known decomposers, bacteria include vast majority of decomposers who largely go unseen in ecosystems. It's been estimated that the biomass of all the bacteria on the planet far exceeds the larger and more easily observed eukaryotes, such as plants and animals.

Autotrophs form a direct link between geology and biology as they use energy to make nutrients available to ecosystems. For example, photosynthesis requires energy to make carbon available to living organisms and nitrogen fixing bacteria require energy to make nitrogen available to plants. Without energy, carbon would quickly become locked up in carbon dioxide and inorganic rocks. Nitrogen would return to the atmosphere as nitrogen gas (N_2), which is just as unusable to most life as carbon dioxide. Other elements and minerals would also quickly become unavailable to life as well. Without energy, life would quickly perish as chemical cycling would grind to a halt. Eventually the Earth would decay towards a stable equilibrium where no living organism could survive.

When plants use energy to fix carbon dioxide, they move carbon into an ecosystem by adding biomass or organic material. Known as primary productivity, it is amount of organic mass added per unit area over a specific time. To understand primary productivity, imagine your lawn: during the summer your grass grows, if you grow one kilogram of grass in a square meter over a year then your net primary productivity would be $1kg/m^2$ -Yr.

Several factors increase or decrease primary productivity, water and energy being the most important. Returning to our lawn example, the more you water the lawn in the summer, the faster the grass grows, meaning it has higher primary productivity. However, in the early spring when it's cooler outside the grass doesn't grow as fast compared to the hot summer months, provided it has plenty of water. Based on this observation, we know that primary productivity is directly related to process called evapotranspiration, which is the amount of water emitted from plants. High rates of evapotranspiration result from plenty of water availability, abundant sunshine, and nutrients required by plants to grow.

In general, moisture, energy, and nutrient availability determine the primary productivity of ecosystems. The amount of energy available to ecosystems varies across the planet. Areas near the equator continually receive abundant energy from the sun throughout the year and have the highest rates of primary productivity. As you move north or southward, seasonal variation increases in sunlight due to the tilt of the Earth on its axis as it revolves around the sun. This tilt makes sunlight come in at an angle at higher latitudes and therefore has to pass through more of the atmosphere. So even when it is daylight in winter, the sunlight doesn't deliver the same amount of energy as the direct sunlight at the equator. We witness this every day, it is hottest at noon when the sun is overhead and cooler in the evening and mornings when the sun is low on the horizon. As a result, northern and southern latitudes, on average, have lower primary productivity simply because there is less energy available, especially in the winter when temperatures fall below freezing in large regions of the north.

Terrestrial ecosystems also vary greatly in productivity based on water availability. Deserts receive plenty of energy with abundant sunshine, but water is scarce. As a result, primary productivity is very low. Worldwide, ecosystems with the highest annual primary productivity are also among the most biologically diverse, whereas those with the lowest primary productivity are the least diverse. This phenomenon is easily observed when you map patterns of diversity against net primary productivity.

The interplay between nutrient and water availability along with energy inputs determine the overall productivity and diversity of ecosystems. This goes for both terrestrial and aquatic ecosystems. Although, aquatic ecosystems are sensitive to changes in nutrient levels, especially nitrogen and phosphorous concentrations. In freshwater and marine ecosystems, clear water is associated with low levels of nitrogen and phosphorous, which limit the abundance of photosynthetic organisms. In fact, large regions of the oceans have really low primary productivity despite abundant sunshine and water, in this case, nutrients limit productivity. When nutrient levels do rise, algae and phytoplankton quickly use the resource, expanding their numbers, reducing visibility and light penetration.

Surprisingly, in aquatic ecosystems, too many nutrients, especially nitrogen and phosphorous, causes an unhealthy condition called eutrophication, which creates algae blooms. We often associate higher rates of productivity in an ecosystem with higher diversity. Unfortunately, this does not always hold true with aquatic ecosystems. Increased algae growth can actually reduce plant growth, thus degrading aquatic habitats for many fish and invertebrates. The real problem occurs when the phytoplankton die. Because oxygen breaks down their remains, eutrophication causes oxygen levels to drop to the point where animals can no longer survive. This happens where Mississippi River empties into the Gulf of Mexico bringing excess nutrients to the gulf coast causing a dead zone nearly 6,500 square miles each summer. Low oxygen levels can also occur at night when photosynthesis stops producing oxygen and respiration becomes dominate, reducing oxygen to very low levels. When nutrient levels rise on coral reefs, algae can grow directly on top of the corals and kill them.

Ecosystems are Connected
Across the Landscape

At first glance, it may seem that some ecosystems are isolated from each other, especially aquatic and terrestrial ecosystems. The water's edge forms a natural barrier for many organisms, thus separating terrestrial from aquatic ecosystems. The reality is that ecosystems, even aquatic and terrestrial ones, are intricately connected across the landscape. When two different ecosystems meet, the area of transition is called an ecotone. One of my favorite ecotones are sandy beaches where the ocean meats the land. Ecotones can also form when between adjacent terrestrial ecosystems too, such as a field and a forest, or when alpine forests quickly transition to alpine tundra.

Ecotones have unique assemblages of plant and animal communities; in turn, increasing the overall diversity across the landscape. For example, on beaches, shorebirds forage on the algae, plants, and animals washed up on shore, or along a river bank, spiders and predatory insects hang out near the water's edge. You won't find most of these species outside of their specific ecotones.

Ecotones between aquatic and terrestrial environments, undergo regular seasonal variation, making them dynamic. For example, in the north, spring snow melt causes higher flows compared to the summer. When rivers flow at their highest, either from routine spring flooding or from heavy rainfalls, the rivers swell and flood their adjacent wetland areas. The fast-moving water alters the river channel creating new habitats. As the adjacent wetlands become flooded, leaf litter from trees lining the rivers provide nutrients and energy to the river. Later in the summer, the waters recede with reduced flows causing algae production to ramp up due to the clearer water. Each year as algae production ramps up in the summer it provides food for larval insects. Once the insects become adults, they emerge from the water to complete their life cycles on land. Terrestrial predators, including birds, lizards, insects and spiders line up along the shore preying on the emerging insects. These predators may account for 50% to 75% of the biomass of the ecotone formed along the stream bank, far above the typical 1% to 10% you would find in most other ecosystems.

Inputs of nutrients and energy change throughout the year, further increasing diversity as communities shift to take advantage of the resources. Stream insect communities are quite diverse, including various species of mayflies, caddisflies, stoneflies, and midges. Some species graze algae, others shred leaves, some live in fast-moving water while others live in slower backwater habitats. Depending on the time of year, the species change in response to different resources.

Flows of nutrients and energy between ecosystems quickly gets complicated. To simplify how it works, I'll use Cutthroat Trout as an example. These fish eat mostly insects and after looking at their gut content, they aren't too picky. They seem to like whatever is in front of them or easy to eat, I know trout fishermen most likely disagree with me here. Let's start with the summer when terrestrial production is in full swing and insect populations become abundant. Many terrestrial insects meet an untimely end when they fall into a stream only to be eaten by a trout. In this case, terrestrial production provides nutrients and energy to the stream, specifically the trout. During the winter, aquatic insects graze algae from rocks, and trout eating these insect larvae depend on aquatic production. This means that trout rely on both aquatic and terrestrial production.

Additionally, the communities of mayflies, caddisflies, and stonelfies change in response to what's available, during the summer species grazing algae from rocks become common, and the shredders relying on leaf fall become abundant in the fall and winter. In the spring and summer when insects with aquatic larvae emerge as adults, they provide an input of resources to the shoreline, where they are eaten by birds, reptiles, and other arthropods. In this case, aquatic production moves to the terrestrial environment acting as a subsidy and increasing diversity. It just goes to show that terrestrial and aquatic ecosystems become intricately connected through the flow of nutrients and energy.

Like many people, I like to take long walks on the beach and poke at dead things with a stick, you never know what you may find. The transition between marine and terrestrial environments include, sandy beaches, mangroves, and estuaries, each one harboring unique communities. I personally enjoy sandy beaches because it's easy to find unique animals, plants, and

Mangrove roots provide habitats for many species found nowhere else

algae washed up on the shore. Along the gulf coast, sea grass routinely washes up providing an input of nutrients, energy, and habitat for small animals on the beach. A quick inspection of seaweed will reveal small mollusks, crustaceans, worms, and other animals which in turn, are eaten by shorebirds and insects. Therefore, a sandy beach with almost no primary productivity from its own plant growth host diverse communities of invertebrates and shorebirds by relying on marine inputs.

Not everything that arrives on a beach from the ocean is dead. On the east coast of North America and the Gulf Coast during the spring high tides, ancient looking horseshoe crabs crawl up to the edge of the surf by the tens of thousands to lay their eggs. Their egg laying event is so predictable that some species of shorebirds, including Red Knots time their spring migration to take advantage of easy meals from all the eggs being laid.

Mangrove forests are another unique habitat in tropical regions on the world. In the New World, there are three species of unrelated trees called mangroves. Each species has unique adaptations allowing them to grow in or near marine environments. For example, Red mangroves use prop roots and roots growing down from their trunk to prop the tree above the water. Their roots also provide habitats for many species of fish, and the roots themselves provide a place for algae, sponges, mollusks, and tunicates to grow.

Ecosystems are Dynamic

Far from being static and unchanging, ecosystems constantly change over time. Changes can be rapid, slow, predictable, or unpredictable. Any change in climate, alterations in disturbance regimes, such as fire and flooding, or the spread of invasive species all cause ecosystems to change. Ecosystems also respond to predictable seasonal changes as well.

Predictable seasonal variations between winter and summer in temperate ecosystems cause seasonal changes in primary productivity. During the winter months, deciduous trees lose their leaves to avoid freezing temperatures, basically shutting down all plant growth. Once it warms in the spring, the trees put out new leaves and begin to photosynthesize and grow. The seasonal changes in primary productivity are enormous, ranging from nearly zero in the winter to rivaling a tropical rainforest.

Ecosystems also change over longer periods of time. After the last ice age, the glaciers retreated across North America leaving mostly barren ground. Slowly, a process called succession took place where ecosystems transitioned from bear ground, to small low-lying plants, to small bushes, and eventually to forests with tall trees. Succession is common, it happens to abandoned farms where fields eventually become forested.

When new oceanic islands form, plants and animals colonize the new habitat and barren islands eventually become forested. On islands, the colonizing species usually come from the nearest mainland.

Changes in climate and disturbance regimes also causes ecosystems to change. Some changes take place over millions of years as the continents wander across the Earth. It's hard to believe how much places can change. Millions of years ago, Antarctica was forested, but as it drifted over the South Pole, it cooled and eventually froze over. Today, the continent is mostly covered in ice, with no forests, and tundra confined to the northernmost areas.

Many ecosystems, including the plants and animals rely on regular disturbance, Unfortunately, we are changing disturbance regimes through climate change and by specific activities like fire suppression and damming rivers, which can be detrimental to the wildlife. Climate change is already causing communities to change worldwide. Nowhere is this more evident than in the southwest from New Mexico to California where the region is experiencing a prolonged multi-year drought. In New Mexico, the foothill ecosystems near the Rocky Mountains were once dominated by pinyon pine and juniper trees. After years of intense drought, the vast majority of the pinyon pines in this area died, leaving mostly juniper trees. In the mountains of southern Colorado, fir trees and aspen trees are dying from a combination of drought and bark-beetle infestations.

In California, severe long-term drought is killing trees, causing more forest fires and changing the composition of the communities from wooded forest to scrublands. If normal amounts of rain and snow do not return, millions of acres of forest will be replaced by more drought tolerant shrubland and grassland communities.

The change from forest to shrublands may not take hundreds of years, but could happen in an instant from forest fires. Routine fire is a natural disturbance in many forests where it removes biomass and recycles nutrients. However, continuous drought combined with higher temperatures driven by climate change, has caused intense forest fires, killing large stands of trees in a few days. Under normal climate conditions, these forests would eventually return, although it may take decades. Unfortunately, if precipitation remains low, then forests will not return, and hardier drought resistant shrub-like vegetation or grasslands will replace the forests.

Changes in natural fire regimes have been drastically changed through fire suppression, infrequent burning, or burning at the wrong time. In the southeast where people excluded fire for decades, less diverse hardwood communities replaced the more diverse Longleaf Pine ecosystems. The succession from pines to hardwoods has been slow, taking decades. But, throughout the southeast, large remnant Longleaf pines, hundreds of years old, live mixed in with much younger hardwoods, such as oaks and hickories. Unfortunately, these large pines can't replace themselves as the young pines need lots of sunshine and do not grow in the dense understory of a hardwood hammock.

What most people don't realize is that the succession to a hardwood forest actually reduces diversity. In the west, the lack of fire in the western forests has led to more trees per acre. While this may sound like a good thing, it too lowers diversity by preventing sun loving wildflowers from growing. Unfortunately, when fire does occur, it is more likely to burn hotter, killing more trees.

Sometimes, changes in the timing of wildfires causes problems too. Returning to the Longleaf pine ecosystems; historically, most of the natural fires occurred every 1-3 years during the spring and summer. However, current management practices on some of our national forests burn the pine forests at longer intervals and during the winter months. Although, fire is still being used to manage this forest, the unnatural timing of the fires from mismanagement due to a lack of using science will cause the loss of healthy forest. The problems become two-fold. First, the longer time between burns allows more fuel buildup and hotter fires that kill young pine trees. Second, burning in the winter months promotes the growth saw palmetto which displaces the native wildflowers, leading to an overall loss of diversity.

Fire is not the only type of disturbance being altered by humans. Many rivers have predictable seasons of high and low flow. Additionally, the plants and animals that live along these rivers depend on the regular timing of the flows. Throughout the southwest, riparian areas were once dominated by large cottonwood trees. The loss of spring flooding due to extensive damming and agricultural use led to the spread of non-native salt cedar trees (*Tamarisk*), further preventing the native cottonwood's ability to replace themselves.

Without the return of natural flow variability to our rivers, riparian areas, including the famous Bosque along the Rio Grande River, is like the living dead. The large cottonwood trees will slowly die over the next few decades only to be replaced by salt cedars and Russian olives (*Elaeagnus angustifolia*). In this case the spread of invasive species is tied to changes in water flows, which drives the succession to new dominant plants and animals living along rivers and their riparian areas.

The spread of invasive species can completely change an ecosystem making it almost unrecognizable. Once established, invasive species displace the native species, causing shifts in the communities, often accompanied with further losses of diversity. One hundred years ago, the spread of a fungus called *Cryphonectria parasitica* caused the chestnut blight that wiped out the American chestnut tree (*Castanea dentata*), which was the dominant tree of the entire Eastern forest. The loss caused a major restructuring in the canopy and directly caused the extinction of several animals. In southern Florida, *Melaleuca* trees (*Melaleuca quinuenervia*) were intentionally spread throughout the everglades in the 1920s. Over the past 90 years they grew into large trees forming species poor-forest where once thriving wetland marshes existed. In aquatic communities, the rise of nutrients helped to spread *Hydrilla* and other invasive plants into our lakes and rivers. Once established, *Hydrilla* displaces the native vegetation and alters fish communities, most of the time skewing them to much smaller fish, much the chagrin of local fishermen.

Similar Climates and Ecosystems
Form a Biome

Worldwide, similarities in climate and ecosystem processes create similar biomes. Biomes are large regions with common characteristics such as rates of primary productivity and patterns of diversity. Although similar biomes on different continents share very few species, those species are similar in their ability to survive in similar environments. Take a tropical rainforest for example, the nearly constant sun creates hot and humid environments with intense competition between species. Many types of temperate trees found in North America could probably live in a tropical rainforest, but they are not found there because they would be out-competed by fast growing species that don't have to shed their leaves for part of the year.

In this section, I will cover seven terrestrial biomes; tropical rain forests, temperate forests, coniferous forests, savannas, grasslands, tundra, and deserts. Not everyone agrees on the same biomes, for example some ecologists recognize different types of biomes, such as chaparral.

Tropical Rainforests

Located near the equator, tropical rain forests have the highest rates of annual transpiration (water coming from plants), the highest rates of primary productivity, and are among the most diverse ecosystems on the planet. Every day for twelve hours, all year long, sunlight reaches the tropics. In these regions, hot air rises and it cools causing the water to condense and fall back as rain. Therefore, tropical rain forests receive plenty of rain and combined with nearly constant temperatures, they have a continuous growing season lasting all year. Coinciding the with the large energy inputs, the rates of evolution and speciation are also quite high.

Primary productivity in rain forest is over twice the rate as in temperate forests of the eastern U.S. The diversity in the tropics is staggering. Take the small country of Costa Rica, at 19,700 square miles it is smaller than West Virginia and only 1/6 the size of New Mexico. Yet, it has over 800 species of birds, more than twice what is found in New Mexico and roughly equal to the number of birds found in all the United States with an area of 9.6 million square miles. Throughout the tropics, tree diversity is also quite high, in regions of the Amazon rain forest in Brazil, over 300 species of trees have been identified in one acre (approximately 43,000 square feet). Once again, in all North America, there are about 1,000 species of trees.

To say that estimating insect diversity in the tropics is difficult, would be an understatement. There could be upwards of a million of unknown insect species in the tropics alone. Recent studies in Ecuador found over

100,000 species per acre solely in the canopy of a virgin forest. To put that in perspective, the Great Smoky Mountains National Park in North Carolina and located in a temperate biome has an estimated 50,000-80,000 species in its 522,000 acres or 816 square miles.

Tropical rainforests also have some of the highest rates of evolution on the planet. Intense competition among species combined with a year-long growing season, and lots of energy have been cited as reasons for the rapid rates of evolution. Some research has also indicated that mutualistic symbiotic relationships, or the cooperation between organisms could also be a major factor driving the rapid rates of evolution explaining the incredible species diversity of the tropics.

Not everyone totally agrees in how to define a biome. Along the northwest Pacific coast of North America is another type of rainforests, dominated by large conifers These regions being farther north receive much less energy, but plenty of rainfall. They are not as diverse as the tropical rainforest, but are home to Redwoods and Sequoias, some of the largest trees on the Earth. Some scientists classify this rainforest in northern latitudes as temperate rainforests to separate them from the tropical rainforests.

Temperate Forests

Temperate forests are found at mid-latitudes, mostly in the northern hemisphere. In the summer, they strongly resemble tropical rainforest in being hot and humid with high rates of primary productivity. These ecosystems are mostly dominated by large broadleaf deciduous trees (trees that lose their leaves on a seasonal basis) and temperate conifers including pines and hemlock. During the summer, they are home to numerous song birds and insects are abundant. By late fall, the temperatures begin to drop, and the deciduous trees begin to lose their leaves greatly reducing primary productivity. Many insects lay their eggs and die at the end of the summer season, while others find places to safely overwinter. Starting in the late summer, nearly 5 billion songbirds migrate southward to the tropics to avoid the winter. Temperate forests can be quite diverse in the summer, but they all undergo the seasonal reductions during the cold winters.

Northern Coniferous Forests

The northern coniferous forests form the largest continuous forests in the world as it stretches across Europe, Asia, and North America in Alaska and Canada. During the last ice age, these forests in North America were connected to Asia and today they are separated by the narrow Bering Sea. Cold tolerant evergreen conifers that keep their needles all winter are dominate. The trees are narrow with short branches to prevent snow from accumulating and breaking them in the winter. They also have other adaptations to prevent the needles from freezing in the cold winters.

Summer is short lived, lasting only a few months when temperatures are relatively warm, and the region comes alive with numerous insects, songbirds, and active mammals mating and storing fat for the long winters. The northern coniferous forests has a large standing crop despite their low primary productivity. Standing crop is defined as the amount of biomass in a certain area. In coniferous forests the trees may be less than a hundred feet tall, but could be hundreds of years old, adding only a little biomass each year during their short growing season.

Tropical Savannas and Temperate Grasslands

Grasses dominate the tropical savannas and temperate grasslands. Trees and other woody vegetation are sparse in these regions for both biotic and abiotic reasons. These biomes typically have a growing and a non-growing season, although the cause of the two seasons is for different reasons. Savannas are found in tropical regions where rainfall determines seasonal variability. The dry season limits productivity and frequent fires combined with herbivory remove biomass. In temperate grasslands cold winters reduce productivity. Because savannas remain warm all year long, compared to temperate grasslands, some ecologists consider them to be different biomes. Both biomes rely on grazing and fire to maintain their diversity by preventing encroachment of woody vegetation.

Temperate grasslands and tropical savannas lack a large standing crop even though primary productivity can be quite high at times. To compare, coniferous forests to the north have a much higher standing crop, but their primary productivity is much lower. It turns out that disturbance plays an important role in maintaining grasslands and savannas. Routine fire every few years combined with intense grazing from mammals removes large amounts of biomass and helps to recycle to the nutrients. Many grassland ecosystems would become much woodier if not for the fire and grazing pressure preventing the encroachment of trees and other woody vegetation.

Tundra

To the far north or above the tree line of tall mountains is the tundra, a biome dominated by low growing plants, lichens, and mosses. The growing season is very short, restricted to only a few months in the summer when the snow melts. Restricted by the brutal cold of the winter months, trees and other woody vegetation are greatly limited, or absent all together.

Arctic tundra is located north of the tree line and much of it lies above the Arctic Circle, where the sun shines continuously for 24 hours every day during the summer. Shorebirds routinely migrate to breed in the Arctic tundra, only to leave once the temperatures begin to drop as the days become shorter. In the summer, only the first few inches of the ground thaw, the rest remains permanently frozen and is called permafrost, which is a major factor limiting the spread of trees northward. Over thousands of years, plant material has accumulated as frozen peat, in some cases, the peat can be thousands of feet thick. Because it is frozen, bacteria and other decomposers cannot break down the organic material, so it slowly accumulates forming large peat deposits.

In the large mountain ranges tall peaks extend beyond the limits of trees, quickly transitioning to tundra. It's commonly called alpine tundra to differentiate it between the Arctic tundra to the north. Once again, the tree line is mostly determined by cold temperatures. As you go up a mountain, the temperature drops about 3.5^0 F for each 1,000 feet in elevation gain. In New Mexico, the tree line is about 11,500 feet, which means the average temperature at this elevation is 38^0 F colder than if you were at sea level. However, the tree line becomes progressively lower as you go northward. Alpine tundra typically lacks the large accumulations of peat found in the Arctic Tundra, but still has mosses, lichens, and wildflowers that bloom and seed before the cold winters.

Deserts

Deserts, known for being dry, may get less than 10-inches of rain per year. They cover nearly 20% of land's surface and are on every continent, with the majority located around 30-degrees latitude. Large atmospheric circulation patterns are mostly responsible for the distribution of deserts. However, local conditions, such as the presence of mountain ranges are also important for creating deserts.

We are most familiar with hot deserts, including the Sahara Desert of Northern Africa, or the Sonora Desert in the U.S. Plants and animals living in deserts must be adapted to the hot dry environments, especially during the summer months. The harsh conditions created by high temperatures and a lack of precipitation and moisture make living difficult for most organisms to survive. Water availability limits plant productivity resulting in scarce vegetation with few large mammals.

Four deserts, the Sonora, Chihuahua, Mojave, and the Great Basin Desert, dominate the American southwest. In New Mexico, the Chihuahua Desert occurs in the southern third of the state. It is a relatively mild hot desert due to its higher elevation. It receives about 8-16 inches of rain per year, mostly in July through September from Monsoonal rains. The dominant plant is the nearly ubiquitous creosote bush, but there are also many species of cacti. Arizona is home to the Sonora Desert famous for its large columnar cacti. Parts of western Arizona and eastern California form the Mojave Desert famous for its Joshua Trees. The Great Basin Desert lies further north and is considered a cold desert

If precipitation defines a desert, then regions of Antarctica or the Great Basin Desert of Nevada and Utah would be cold deserts. Most images of Antarctica include vast expanses of ice. However, mountains and extreme cold prevent moisture from reaching certain areas creating some of the driest places on the Earth. In fact these regions are so dry, that any snow evaporates faster than it is replaced creating dry valleys like McMurdo dry valleys.

Marine Ecosystems

Marine ecosystems are the most common ecosystems on the Earth. Oceans cover approximately 70% of our planet with an average depth of nearly 2.4 miles. Under good conditions, light only penetrates to about 600-700 feet in the open ocean. Therefore, the two most common habitats on the planet are the aphotic zone and the abyssal plain. Many of the animals living here, never see sunlight. These deep-sea organisms can appear quite alien to us. Some make their own light through bioluminescence to attract prey which they catch with their enormous mouths full of long teeth.

These deep see communities have very few if any autotrophs. Instead, they receive the bulk of their energy as planktonic snow, the slow but steady trickle of dead organisms that float down from the surface to the depths of the ocean. Some deep-sea organisms make nightly migrations to within a few hundred feet of the surface to take advantage of more resources. In recent years, it was discovered that whale carcasses form important habitats and provide resources to the abyssal plain. In fact, one study of a single whale carcass over a two-year period discovered new species thought to depend solely on dead whales sinking to the bottom of the ocean.

Thermal vents are found in the depths of the oceans where powerful geological forces push the continents apart. The ecosystems surrounding the thermal vents are teaming with life, yet not a single organism depends on energy from sunlight. These unique communities were discovered in the 1970s and represented an entirely unknown and unique ecosystem dependent solely on chemosynthesis. Here, small autotrophic bacteria extract energy directly from the minerals spewing out of the vents and serve as the base of the food web. Hot vent communities may only last a few decades before they become inactive and the inhabitants must migrate to new active sites.

Perhaps the exact opposite of a thermal vent are the coral reefs. Found in warm shallow waters, these are among the most diverse ecosystems in the world. Although, they occupy less than one percent of the ocean's surface, they may account for nearly half of its diversity. Coral reefs are dependent on the corals that build structure by excreting calcium carbonate from the ocean water.

Kelp forest are very productive ecosystems found mostly near the coast from temperate to polar climates around the world. Kelp is a general term for several species of large algae where some can grow over 100 feet in length creating a canopy. These large canopies provide important habitat for many other species of fish and invertebrates.

Also found in shallow waters close to shore are seagrass beds dominated by flowering plants that superficially resemble grasses because of their long narrow leaves. Many fish living on coral reefs visit seagrass beds to feed at night. These ecosystems have very high rates of primary production and serve as nursery grounds for many species, including several popular sport fish and bay scallops.

My favorite marine ecosystems are the coral reefs. With productivity and diversity rivaling rainforests, these systems account for almost half of the marine diversity, yet less than 1% of the ocean's surface. To live in the nutrient poor waters, corals rely on a symbiotic relationship with their zooxanthellae, which makes sugars through photosynthesis. Many corals secrete calcium carbonate, a rock which provides additional structure providing more places for other animals to live. Unfortunately, coral reefs are among the most imperiled ecosystems in the world

Chapter 13

Climate Change

Science is not political,
If I tell you that more carbon dioxide in the atmosphere
will raise the Earth's temperature,
it is borne out of observations and experimentation
that can be repeated.
It is not an ideological assertion based on a belief system.

Introduction

Life is connected to the Earth. For over 3.8 billion years, life has been altering the composition of the atmosphere, the oceans, and rocks on its surface. When life first emerged, the atmosphere had very little oxygen, perhaps less than a thousandth of a percent. Instead, it was likely dominated by carbon dioxide and nitrogen. The air we breathe today is about 22% oxygen, a byproduct of photosynthesis, a process of life. In fact, if scientists were to discover another planet with an oxygen rich atmosphere, it would be a smoking gun for life.

As the composition of gases in the atmosphere changed over time, so has the climate. Before I paint a picture that climate and the atmosphere are solely influenced by life, other factors have greatly changed the climate, including volcanism, the position of the continents, and changes in the Earth's orbit. But one thing is certain, when the climate changes, it has an enormous impact on life. Species go extinct, new species evolve, entire ecosystems collapse while new ones emerge.

For billions of years, the Earth's climate has changed between hothouse and icehouse climates. Most changes occurred slowly over the course of millions of years at the pace of continents moving across the globe, while other periods climate change include rapid glaciation or melting that take place over centuries. One important point to remember; in the past, rapid climate change caused several mass extinction events.

Once again, the Earth's climate is rapidly changing. Although, this time it's not from natural causes, but from the burning of fossil fuels releasing billions of tons of carbon dioxide into the atmosphere every year, an unprecedented rate not seen in the Earth's past. We know this from an enormous amount of evidence including the loss of glaciers and sea ice, rising sea levels, and record temperatures, all clearly showing a warming planet. By refuting natural causes of climate change, we know with a high degree of certainty that we are causing global climate change, which will have a huge impact to all life and our civilization.

An Introduction to Climate

The Earth's climate is complicated with many factors causing it to be stable or to change, and at times drastically. Before diving into climate, let's begin by clarifying the difference between weather and climate. Weather is the daily activity of the atmosphere. Weather is what is happening right now; is it rainy, snowy, sunny, or windy, hot, or cold. And weather can change rapidly, an oncoming thunderstorm can drop the temperature by 10-20°F in less than an hour. Many variables influence local weather events making it difficult to forecast beyond few days with any accuracy.

Climate, on the other hand, is the average weather you would expect for a region at a specific time. When you watch a weather forecast and the meteorologist claims the high temperature for the day is above average, they are comparing that day to a 30-year average. The climate of any one region is determined by a complex mix of local geology, such as the presence of mountain ranges, the position of the continents, the latitude of a region, large scale global circulation patterns, history, and of course the composition of gases in our atmosphere. Tropical rainforests experience a climate with constant warmth, and average daily highs and lows may fluctuate less than a few degrees throughout the year. In contrast, temperate forests may experience a 40°F difference in average seasonal temperatures.

The atmosphere blankets the Earth and together with ocean currents they redistribute energy around the planet. If it weren't for the atmosphere, daytime temperatures in the tropics would soar hundreds of degrees during the day only to plummet to -200°F at night. The poles would be frigid with temperatures lower than -200°F. Almost certainly, life would have a difficult time existing without the atmosphere and oceans to redistribute the incoming solar radiation and mitigate day and night temperature differences by acting as a natural greenhouse.

The composition of gases in the atmosphere is a major factor influencing how much heat is retained. A mixture of gases forms the air we breathe where the current composition of our atmosphere is about 78% nitrogen, 21% oxygen, and about 1% trace gases including carbon dioxide and methane along with some water vapor. Water vapor, carbon dioxide, and methane are the most abundant greenhouse gases because they retain heat in the atmosphere. For example, night time temperatures remain higher on cloudy nights, or when the air is humid. The water vapor in the atmosphere retains heat from the day and prevents it from escaping into space at night. Carbon dioxide and methane also trap incoming solar radiation. Even though they are trace gases, small changes lead to large changes in the climate. We are currently witnessing this phenomenon right now as increased carbon dioxide in the atmosphere is causing the current rise in global temperatures.

A History of Climate Change

Imagine a time so cold that glaciers covered the planet making it look like a giant snowball, jump ahead millions of years to a time when the near arctic had tropical climates and there were no glaciers or permanent ice anywhere on the planet. The history of climate on the Earth has been one of change, throughout Earth's long history, the climate has experienced episodes of extreme ice ages to times of warmth known as a hothouse. Sometimes, the climate would remain stable for millions of years, whereas other times the climate would change rapidly over just a few thousand or even hundreds of years.

Reconstructing past climates becomes more challenging the further back in time you go. The movement of continents and the constant erosion of the surface as rocks are recycled remove evidence for Earth's deep past. Today, only a few places have rocks older than about 3.5 billion years, so we know very little of the ancient Earth. We use three types of evidence to reconstruct ancient climates including geological, chemical, and paleontological records. Closer to modern times, we use additional sources of information including ice cores, sediment cores, and tree ring data accurately reconstruct the past climate.

For much of the Earth's history, atmospheric, geological, and life's processes have been intricately connected, they do not exist in isolation of each other. The Earth's climate is strongly influenced by the composition of gases in the atmosphere, which in turn, is a product of geological and living processes. Prior to the origins of life, the atmosphere was influenced by geological processes such as intense volcanism releasing carbon dioxide into the atmosphere and rock weathering that removed carbon dioxide. Based on our best evidence, the atmosphere 4 billion years ago was an alien atmosphere, a mix of methane, carbon dioxide, nitrogen, and water. It was unable to support animal life because it lacked oxygen. Carbon dioxide and methane, were likely much more prevalent, which was a good thing because the sun was about 25% cooler than it is today. It is thought that the high levels of these two greenhouse gases, along with water vapor, kept the Earth warm and prevented the oceans from freezing.

Recent theories suggest that both living organisms and geological processes worked together to stabilize the atmosphere early on, preventing runaway greenhouses like what happened to Venus, or preventing the loss of our oceans like what happened to Mars. I should point out that Mars is dry due to its smaller size and lack of a magnetic field to protect it from the solar wind. Although the Earth's climate has not gone to quite the extremes in climate as our two nearest planetary neighbors, the Earth's climate has undergone major swings when global temperatures have fluctuated by nearly 27°F (15°C). At the cold end during severe ice ages, global average temperatures were about 50°F (10°C) and during warmer periods, global average temperatures may have spiked above 77°F (25°C).

Although, 77°F may seem warm, remember, this would be the global average. Temperatures in the tropics could have easily reached 120°F, an unbearable temperature for most living things. Also, other findings show that ocean temperatures on the surface may have reached about 100°F, making it almost impossible for most animals to survive in such extreme environments. In comparison, the current average global temperature is about 58.7°F (14.8°C), and ocean surface temperatures in the tropics range about 80-83°F, perfect for high rates of diversity.

The reason why we care about climate and climate change is because each time the climate changes, it has had a profound impact on the diversity of life and its future evolution. Rapid climate change has caused mass extinctions while milder, more stable climates have led to rapid speciation. Based on our evidence there have been at least five major ice ages lasting from a few million years to hundreds of millions of years.

The Proterozoic Ice Ages

Evidence for life causing the first major change in climate dates to at least 2.4 billion years ago in the Proterozoic Eon. Ancient cyanobacteria continually pumped oxygen into the atmosphere for over a billion years, slowly changing the mixture of gases in the atmosphere causing the Great Oxygen Event. The oxygen produced from photosynthesis removed methane, a greenhouse gas, from the atmosphere. Over time, the loss of methane cooled the Earth and it entered an ice age lasting nearly 300 million years. Luckily, continual volcanism released carbon dioxide into the atmosphere, eventually warming the climate and ending the Earth's longest ice age.

Once again, about 720 to 635 million years ago, the Earth entered another ice age called the Cryogenian Period. This ice age may have been so severe that glaciers on land reached the tropics and sea ice extended almost to the equator covering most of the planet. Some think the glaciation on land and sea was so drastic they call it the "Snowball Earth". Because we don't have much evidence from this time period, not everyone agrees with these theories questioning the extent of the glaciation and instead call it the "Slushball Earth". However, evidence suggest that carbon dioxide levels were quite low during this time. So, the question becomes; what removed carbon dioxide from the atmosphere faster than it was put into the atmosphere and how did the Earth get out of this planet-wide 100 million year ice age?

The answer lies in volcanism, active volcanoes pump carbon dioxide into the atmosphere. Luckily, carbon dioxide never built up to catastrophic levels as seen on Venus where surface temperatures are hot enough to melt lead. Part of the reason why that has not happened on Earth is that chemical weathering from rocks and carbon dioxide constantly removes carbon dioxide from the atmosphere. About 700 million years ago the continents drifted near the equator forming a super continent called Rodinia. Being subject to higher rainfall, the continental rocks experienced higher rates of chemical weathering, thus removing carbon dioxide from the atmosphere as it became sequestered into carbonate rocks. The removal of carbon dioxide from the atmosphere lowered the Earth's average temperature causing the spread of glaciers.

Through positive feedback, the glaciers would have continually grown, eventually reaching the tropics. It may have gone something like this: Less carbon dioxide caused colder winters leading to more ice cover. Then, cooler summers resulted in less ice being melted, leading to more ice being formed in the winter. More ice would reflect additional energy back to space, thus further cooling the planet, leading to more ice in a positive feedback cycle leading to a runaway ice age for nearly 100 million years. But, the cold was not to last forever.

During the cryogenian, several factors contributed to the slow rise of carbon dioxide. For one, volcanism never stopped pumping carbon dioxide into the atmosphere. Additionally, photosynthetic organisms are a major force removing carbon dioxide. About 700 million years ago when the Earth was in a runaway ice age with ice covering much of the oceans, the removal of carbon dioxide by photosynthesis was greatly reduced. The glaciers also covered the continents, preventing the chemical weathering of rocks, further slowing the removal of carbon dioxide. Eventually, carbon dioxide in the atmosphere began to accumulate to levels that would trap enough heat to break the grip of ice age.

Eventually the climate warmed and about 600 million years ago the glaciers retreated. In yet another example of a positive feedback the ice age ended; as the glaciers retreated, less energy was reflected to space, rather it was absorbed by the oceans causing the glaciers to further retreat. Eventually, Rodinia began to break apart when the continents drifted southward, which reduced chemical weathering allowing carbon dioxide to continue its slow build up in the atmosphere. By the end of the Proterozoic Eon 542 million years ago, the Earth was quite warm and carbon dioxide levels were 5 times higher than today.

The Phanerozoic: Hothouse - Icehouse

It's been hypothesized that the intense ice age ending 600 million years ago drastically slowed the evolution of life, delaying the appearance of the first animals. Indeed, the start of the Cambrian Period, which is defined by the appearance of abundant animal fossils (specifically the first appearance of trilobites), took place nearly 55 million years after the end of the intense glaciation. For much the last 542 million years of the Phanerozoic Eon, the Earth's climate has gone from warm and wet periods to ice ages much like the one we are in today. But nothing as intense, or as long as the global glaciations during the Proterozoic Eon.

The Phanerozoic Eon remained warm for nearly 100 million years until an ice age gripped the planet nearly 450 million years ago. It's been thought that during this time the continents were in the southern hemisphere and moved over the South Pole causing massive glaciation. The average global temperatures dropped about 18°F (10°C) to an average temperature around 55°F and caused a mass extinction of early sea life. Luckily, this was a short-lived period of glaciation and the glaciers were much less extensive than the previous Snowball Earth. Over time, the continents slowly drifted away from the South Pole and surface temperatures once again returned to their previous levels.

The Devonian Period 400 million years ago was a unique time in the Earth's history. Warm climates and the land masses were once again near the equator leading to rapid diversification of fishes. Some even refer the Devonian Period as the *Age of Fishes*. On land, plants were evolving into the first forests comprised of giant fern trees forming vast swampy forests. Arthropods had colonized the land and primitive insects were evolving flight for the first time.

By 300 million years ago, oxygen levels were much higher than today, accounting for 30% of the atmosphere, allowing arthropods to grow much larger than today. Recall that the oxygen in the atmosphere comes from photosynthesis and is removed by rock weathering. During the Carboniferous, the Earth was covered in vast forests pumping lots of oxygen into the atmosphere. These forests covered the surface, preventing rock weathering, as a result, both oxygen and carbon dioxide levels soared.

The spread of ancient forests may have initially caused carbon dioxide levels to rise, but they also caused them to drop later on. Over millions of years, the ancient forests slowly removed carbon dioxide from the atmosphere. As the plants died and fell into the swamps, bacteria were unable to fully break down the organic matter resulting in large deposits of coal and oil. Overtime, carbon dioxide levels fell from 2,000 ppm to about 250 ppm and the world once again entered an ice age. It was followed by a minor extinction event, although not one of the Big five mass extinction events. The slow change in climate spurred the evolution of seed plants and amniotes in response to the cooler, drier climates.

About 251 million years ago, near the end of the Paleozoic Era, carbon dioxide levels and temperatures rapidly increased from extensive volcanism in present day Siberia. The intense volcanism not only changed the atmosphere, it made the oceans more acidic, causing the largest mass extinction of the Phanerozoic Eon. It has been estimated that 80-95% of all life perished in less than a million years, a very short time in the geological record.

The End Permian extinction marked the end of the almost 300 million years of the Paleozoic Era and ushered in the Mesozoic Era. The start of the Mesozoic Era was very hot, perhaps the hottest the Earth had ever been with some evidence suggesting that surface ocean temperatures reached 104°F. After about 20 million years, average global temperatures cooled, but remained high for much of the Mesozoic Era, known for the reign of the dinosaurs. However, beginning 100 million years ago carbon dioxide levels began to slowly decline. By sometime around 35 million years ago, the Earth's climate began to cool and dry once again.

Eventually polar ice caps began to form about 34 million years ago. By the start of the Pleistocene Epoch 2.6 million years ago, the Earth, once again, entered another ice age that we are still in today. Although some scientists think that we have been in an ice age for 34 million years, placing the start with the freezing of Antarctica. Antarctica has not always been covered with large glaciers, prior to 34 million years ago, it was warmer and forested.

For the last 450,000 years, and perhaps 2 million years, the Earth has oscillated between periods of extensive glaciation and periods of reduced glaciation known as interglacials. Currently, we have been in an interglacial for the last 10,000 years. Several of the interglacial periods have been warmer by approximately 3.9°F (2°C). This temperature difference may not seem like much, but it corresponds with sea levels about 20 feet higher than today. When you look at the cyclic nature of climate change over the past 450,000 years, the tight relationship between carbon dioxide levels, global temperatures, and sea level becomes apparent. For the last 10,000 years, we have enjoyed a period of relatively stability in the climate that has seen the unprecedented rise in human civilization.

Evidence for Current Climate Change

Science is not political, it depends on making observations, formulating hypotheses, testing hypotheses, and keeping ones that fit the data while discarding ones that do not. Science builds on previous knowledge and changes so that our world view reflects our current state of knowledge. The same holds true with our knowledge of climate change. Let's explore the evidence of modern climate change based on scientific observations. I use the term scientific observations to make it clear that the data is accurate and can be observed or replicated by anyone who's looking.

Believe it or not, the evidence supporting modern climate change dates to Arrhenius in the 1890s. However, I'm going to skip ahead to the late 1990s when the climate controversy really started to become mainstream and heavily politicized when Michael Mann and his colleagues published what became known as the Hockey Stick Model of modern climate change. To reconstruct the past climate, they used proxy data from tree-rings, corals, sediment cores, and ice cores going back about 1000 years. For the past century, they included actual temperature measurements to the model, which were quite revealing of the rapid rise in temperatures in the 20th century.

With the current temperature data included in the model, it became commonly known as the Hockey Stick Model of climate change due to its resemblance of a hockey stick. There are several other conclusions easily drawn from the graph. First, the past climate was relatively stable. There were warmer and cooler periods, although the fluctuations were less than 1°F (0.5°C). Second, the average temperatures for the last 1000 years were below the average temperature set between 1961 and 1990. Third, the average temperature increased nearly 1.25°F (0.7°C) from 1900 to 1998, coinciding with the industrial revolution and the burning of fossil fuels. They also showed that the 1990s were the warmest decade in the last thousand years, and 1998 was the warmest year of the millennium. And lastly, the rapid rise in temperature of the last 100 years corresponded with the rise in carbon dioxide in the atmosphere from the burning of fossil fuels.

Surprisingly, the validity and accuracy of the peer-reviewed graph was questioned by non-scientists, politicians, and industry specialist.

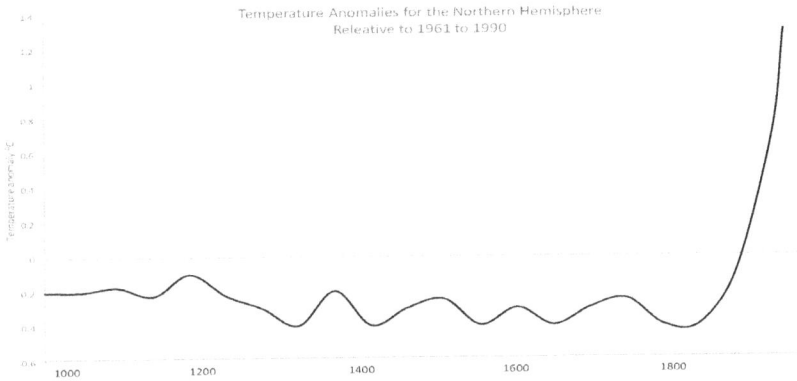

Temperature Anomalies for the Northern Hemisphere
Releative to 1961 to 1990

A graphical representation of the Hockey Stick Model of climate dating back 1,000 years based on proxy data Generated by Mann *et al.* 1998. For approximately 900 years, the climate was relatively stable, but in the last 100 years, average temperatures have risen about 1.3 degrees Celsius. This graph includes the latest temperatures from 2016, to further emphasize the rapid rise in global surface temperatures.

To be blunt, the anti-climate change segment of society attempted to discredit Dr. Mann and his work. However, scientific results are available to everyone so the methods can be repeated and the results can be independently verified. After the political attacks to the Hockey Stick Model and Michael Mann, other scientists set out to verify it by repeating his work.

Within a few years, the work was repeated and expanded only to confirm the rapid temperature rise of the last century. Their findings exonerated Dr. Mann's work, further lending support that Earth's average temperature is rapidly rising. In fact, the rapid rise of the last 100 years is unparalleled in the geological record. It may be the fastest change in climate ever observed.

Since the publication of the Hockey Stick Model, we have had the warmest decade on record from 2001-2010, surpassing the 1990s. The record heat of 1998 has been broken several times, 2014 was the warmest year on record only to be broken again by 2015, and then again in 2016. The record heat of 2015 and 2016 coincided with a super El Nino, which broke the one-week record for warmest sea surface temperatures. July of 2016 was the warmest month ever recorded and

January 2016 broke the record for the largest temperature departure for a month. According to NASA, by 2016, the average global temperature was 2.4°F (1.3°C) above the late 1800s. In less than 20 years, the world almost doubled the amount of warming witnessed from 1900-1999.

In addition to the rise in surface temperatures, the Earth's cryosphere, the frozen part of the planet, has been rapidly declining for the past 50 years. Starting in 1979, NASA sent up satellites to continuously monitor the seasonal variations in Arctic sea ice, providing a much more complete picture than previous records made based on location.

Since these first direct satellite observations, the Arctic sea ice has been rapidly declining at about 13.4% per decade. While 2016 did not set a new record low for Arctic sea ice, it tied 2007 for the second lowest extent in square miles below the 1981-2010 average, or about 40% less area than the late 1970s. In addition to the loss of Arctic sea ice, the Greenland ice sheet also experienced warmer than average temperatures leading to increased melting of the glaciers. Since 2012, ice loss in Antarctica have tripled, losing approximately 241.4 billion tons of ice. In fact, worldwide, most glaciers have been rapidly retreating. Glacier National Park is a case in point. In 1850, the area that forms the park had 150 active glaciers, today it only has 25 active glaciers. In a stable environment, you would expect some glaciers to be retreating, some to be stable, and others to be growing. Instead, we see about 85-90% of all glaciers are retreating worldwide.

In the 1990s, climate models developed by climatologists predicted that warmer climates would cause more severe weather. This intensification of weather patterns would mean intensification of persistent drought, flooding, more intense storms, and an increase in more hurricane and cyclone activity. In the last decade, these predictions have come true in spectacular fashion.

Persistent long-term drought has gripped parts of the west, punctuated by a single wet season, only to have drought return. California has been caught in a long-term persistent drought starting in 2012 and continued through 2018 despite an El Nino and a record 2017 snow pack. There was another intense drought across the Midwest in 2012 greatly reducing crop production. The long-term drought has led to record wildfires in the California, Arizona, and New Mexico burning millions of acres of forests.

It doesn't take fire to kill a forest. In Colorado and New Mexico, warmer temperatures is causing enormous loss of trees from drought stress and the spread of the bark beetle. In 2017, I went snow shoeing at Wolf Creek Pass in southern Colorado only to be shocked by the 50% or more loss of trees at higher elevations from the bark beetle. The problem comes from the combined effect of warmer winter temperatures and reduced snow pack. Warmer winter temperatures don't kill the bark beetles and reduced snow pack leads to drought stress making the trees more susceptible to the beetle. It's a one-two punch for these forests.

Warmer surface temperatures also intensify global patterns of El Nino, drought and flooding where dry places become drier and wet places become wetter. In 2015, the eastern Pacific in the El Nino region recorded its highest temperature ever, surpassing the El Nino of 1997. Surprisingly, it didn't bring record snow to California to end its long-term drought, that happened in 2017. In Australia, and other deserts world wide, more frequent and intense heat waves kill thousands of birds and bats. The number of record-setting dry months has increased by 50% in sub-Saharan Africa. In contrast, the number of months with record high rainfall increased by 25% since 1980 in the eastern U.S.

In an ironic twist, a warmer Arctic causes cold winters in more temperate regions on the northern hemisphere. The Arctic is warming faster than any region on the planet. Historically, a large temperature gradient between the Arctic and lower latitudes created a more or less straight jet stream preventing cold Arctic air from dropping south. The warmer Arctic winters weakens the jet stream allowing it to drop south and with it cold Arctic air moves into large areas with record low temperatures and snow falls. This example goes to show that "common sense knowledge" based on limited knowledge and experience is often wrong, especially in the larger picture. A colder winter in your area does not mean there's no more climate change

Global warming may not cause any one particular storm or hurricane, but it does make them worse. Both hurricane Harvey (Texas 2017) and Michael (Florida 2018) rapidly intensified into major storms. The rapid intensification was a direct result of warmer waters in the Gulf of Mexico.

Hurricanes aren't the only storms getting warmer, large-scale tornado outbreaks and winter storms are becoming more frequent.

Sea levels are rising from the melting of glaciers and warmer waters. Although, sea level rise is not uniform because of ocean currents, the distribution of the land, and some areas are sinking while others are rising. Despite the complexity of sea level rise, satellite measurements, tidal gages, and sediment cores all paint the same picture of accelerating sea level rise. Just ask anyone from South Florida who now faces "sunny day" flooding from high tides that aren't caused by storms. These floods were once rare, but have become common place in low lying areas.

Taken together, the rise in surface temperature, loss of the cryosphere, sea level rise, and the increase in extreme weather events all point to rapid climate change. To put it bluntly, there is no evidence to refute climate change. All the observations around the world clearly show the climate is changing. Based on these observations, climatologists have focused their attention on studying the factors that caused previous climate change and how the Earth will be affected by current and future climate change.

The Natural Causes of Climate Change

For more than a century, scientists reconstructed the past climate to determine the causes of climate change. It's important to do so because by understanding the past, we can understand the present. We know the Earth's climate is a complicated system where small changes can lead to new climate regimes. Natural causes of climate change are also quite varied, they include the movement of continents, periods of intense volcanism, changes in the Earth's orbit, changes in solar output, meteor impacts, and the processes of life like photosynthesis and respiration. And each of these factors can amplify or mitigate each other. But, the take-home message is that many of these factors affect the composition of gases in the atmosphere, which leads to global climate change.

Using proxy data, including tree rings, ice cores, sediment cores, and even corals, we have reconstructed the climate going back about 850,000 years. The results clearly show that our climate has swung between times of maximum and reduced glaciation coinciding with a change of about 9°F (5°C) and carbon dioxide levels fluctuating between a low of 185 ppm to around 300 ppm over periods of hundreds of thousands of years. In addition to the cyclic nature of the climate, the data also showed that warming was quite rapid and lasted for shorter time periods than the ensuing periods of maximum glaciation.

What were the causes driving the cyclic nature of our climate for the past 850,000 years and how could they be affecting us now? For the remaining part of this section, I cover the natural causes of climate change and show how scientists can rule them out as causes for the current climate change.

Milankovitch Cycles

One explanation for the climates' cyclical pattern of the last million years was put forth by the Serbian astronomer Milutin Milankovitch. He described three dominant cycles on scales lasting 26,000 to 100,000 years. Known as Milankovitch cycles, they include eccentricity, axial tilt, and precession.

Eccentricity is the shape of the Earth's orbit around the sun and it changes over time from nearly circular to more elliptical. Each cycle lasts about 100,000 years. Currently, we are at a minimum in the cycle where the Earth's orbit is nearly circular with only a 3% difference between the closest and farthest point from the sun. This means we get about 6% more energy during January. However, at the maximum of the cycle, the difference in energy could 20-30% different.

The tilt of the Earth's axis known as obliquity varies between 22.5° and 24.5° over a period of 41,000 years. The tilt of the Earth's axis causes the different seasons of winter and summer. In the northern hemisphere's summer, the North pole is tilted towards the sun causing warmer temperatures. The tilt of the Earth's axis is currently about 23.5° and is getting smaller. A smaller axial tilt is predicted to promote the growth of ice sheets

Precession is the wobble of the Earth's axis, or changes in the direction of the axis. Today, the North pole points at the North star called Polaris, but thirteen thousand years ago, the North pole pointed at another star. It takes about 23,000 years for this cycle to complete itself.

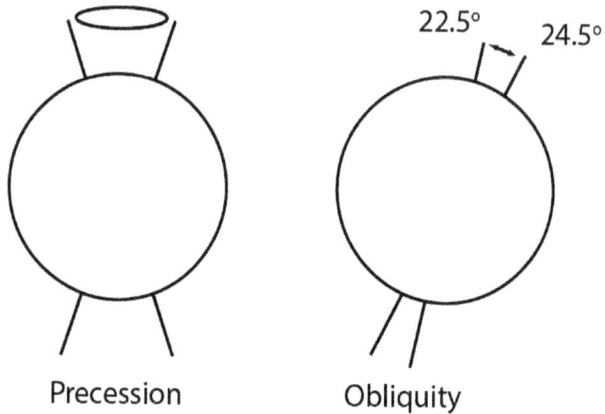

Precession   Obliquity

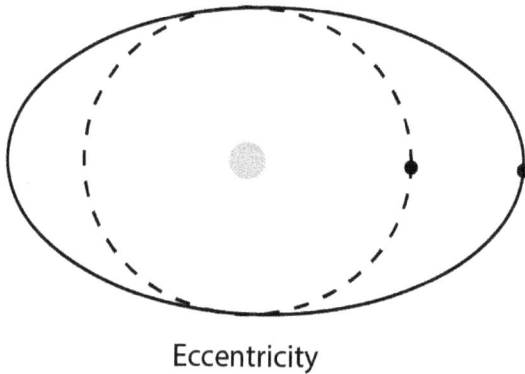

Eccentricity

Milankovitch cycles include the eccentricity of the Earth's orbit, the obliquity, which is the tilt of the axis, and precession which is the wobble on of the axis.

The importance here is that currently, the northern hemisphere points away from the sun when during summer, at the other end of the cycle, it would point towards the sun in summer, making for warmer summers and colder winters.

The interplay of these three cycles affects the global climate. Based on the current Milankovitch cycles, the tilt of the axis is decreasing and combined with a nearly circular orbit, the Earth would naturally be slowly cooling. Therefore, scientists have ruled out Milankovitch cycles as a cause of any current warming trends.

Volcanism

Periods of intense volcanism temporarily cool the planet by emitting large amounts of ash and aerosols into the atmosphere. In 1991, Mt. Pinatubo exploded in the Philippines releasing 20 million tons of sulfur dioxide into the atmosphere temporarily reducing global temperatures by 1°F. However, long-term increases in volcanism release large amounts of carbon dioxide, raising atmospheric concentrations enough to warm the planet. At least two of the last three mass extinctions have been blamed on periods of intensive volcanism, including the End-Permian 251 million years ago and potentially another round of volcanism 201 million years ago, causing the end of the Triassic. There is also evidence of yet another major round of volcanism in modern day India known as the Deccan Traps, which may have been the major factor leading to the extinction of the dinosaurs.

Today, there has not been any major increase in volcanism or geological activity that would lead to greater amounts of carbon dioxide in the atmosphere. We can also analyze the isotopic signatures of carbon dioxide in the atmosphere, which shows that the majority of new CO_2 comes from fossil fuels and not volcanic activity. Once again, we can rule out this natural cause of warming climates.

Meteor Impacts

Large meteor impacts drastically alter the climate in a day. It is believed that a large meteor hit the Earth 65.5 million years ago and caused such a rapid change in climate that it wiped out the dinosaurs. However, more evidence is beginning to support that the meteor impact was preceded by a period of intense volcanism in present day India creating the Deccan Traps that may have caused severe declines in dinosaurs prior to the meteor impact. There have been no recent meteor impacts significant enough to cause any current climate change.

Solar Output

Over the last 4.6 billion years, solar output has continually been increasing. Early in the Earth's history, the sun was less intense, emitting only 70% of the energy it does today. In the far future, over the next few billion years, the sun will continue to increase in brightness as it burns through its hydrogen.

On shorter timescales, the sun undergoes an 11-year solar cycle from periods of low to higher intensity. There may be other longer-term cycles including a 90-year cycle, but they are not as well known. One of these longer cycles may be responsible for the Maunder minimum, which are periods of reduced sunspot activity that correspond to lower solar output. The lower solar output with fewer sunspots corresponded to the Little Ice Age when European and North American temperatures were below average between 1645 and 1715.

Current observations of the sun have shown that we have been in a period of lower solar output that will last until the late 2020s. After that the sun's output will return to normal in the following decades. Based on solar output, the climate should be cooling. Therefore, we can rule solar activity for the current warming we have been observing.

Position of the Continents

The position of the continents can create the conditions for an ice age. When continents block the movement of warm water currents to the poles, they cool off and permanent ice forms. For example, Antarctica sits on top of the south pole and is the coldest place on the planet. The Arctic is mostly landlocked by North America, Europe and Asia preventing the movement of warm water currents into the region. Rodinia may have caused the Snowball Earth 700 million years ago by sitting at the equator, preventing warm water currents moving to the poles. Also, being at the equator, it would have had high rates of rock weathering, acting as a giant carbon dioxide scrubber, removing carbon dioxide leading to cooler temperatures.

The current ice age may have been triggered, in part, by the movement of the continents preventing warm water moving to the poles and allowing them to become colder. Additional intensification of the current ice age came from continental uplifting of the Colorado and Tibetan Plateaus removing large amounts of carbon dioxide from the atmosphere, in a situation like Rodinia. In the last 100 years, the continents have not moved sufficiently to have any warming effect on climate change. Therefore, this has been ruled out as well.

Changes in the Earth's Atmosphere

Our knowledge that carbon dioxide is a greenhouse gas dates back over 125 years ago to the French chemist Svante Arrhenius who predicted that the burning of coal could warm the climate. Greenhouse gases, such as carbon dioxide, methane, and water trap heat and prevent it from escaping back to space. The more greenhouse gases in the atmosphere, the warmer the Earth's climate will be. There are multiple factors that affect the amount of greenhouse gases in the atmosphere. Volcanism can add carbon dioxide to the atmosphere, whereas rock weathering and photosynthesis act to remove carbon dioxide.

In the past 125 years, atmospheric carbon dioxide levels have risen from 285 ppm to over 400 ppm, levels not seen in over 2.2 million years. Additionally, there have been no major changes in volcanic activity or other natural sources causing the rapid increase in carbon dioxide levels.

Although there are natural causes of climate change, the take home message here is that they do not explain the current warming. Often, several factors take place at once to make the Earth hot or send it into an ice age. But most importantly, we know that when the composition of the atmosphere changes, so does the climate. Warmer climates have high levels of carbon dioxide and cooler climates associated with ice ages have low levels of carbon dioxide. And when the climate changes quickly, extinction events are likely to occur.

Humans are the Cause of Modern Climate Change

The cause of modern climate change is well known; the rapid rise of carbon dioxide in the atmosphere from the burning of fossil fuels and to some extent, deforestation is warming the planet. Recall that the French chemists Arrhenius predicted in the 1890s that the burning of fossil fuels would lead to a warmer planet. Prior to the industrial revolution, carbon dioxide levels in the atmosphere were approximately 285 ppm and had been at about that level for the previous 10,000 years. Today, it's a different situation, where they reached over 410 ppm for the first time in more than a million years, and it continues to rapidly rise.

The first daily monitoring of atmospheric carbon dioxide began in the 1950s at the Mauna Loa observatory in Hawaii. A climatologist named Charles Keeling wanted to test whether carbon dioxide levels would decline in the summer and rise in the winter in response to seasonal variations in the northern hemisphere. The thought behind his idea was that photosynthesis in the summer would draw down carbon dioxide levels, but during winter when the trees dropped their leaves, levels would rise again.

He collected data for several years and easily detected the seasonal changes he predicted. However, in science, one discovery often leads to others. Dr. Keeling also noticed that the annual average levels of carbon dioxide were increasing. Today, the same observatory continues taking measurements of carbon dioxide levels. In 2015, the average carbon dioxide levels crossed over 400 ppm, a level that has not been on Earth for over two million years or more. The last time carbon dioxide levels were that high, the oceans were 20 feet higher than they are today.

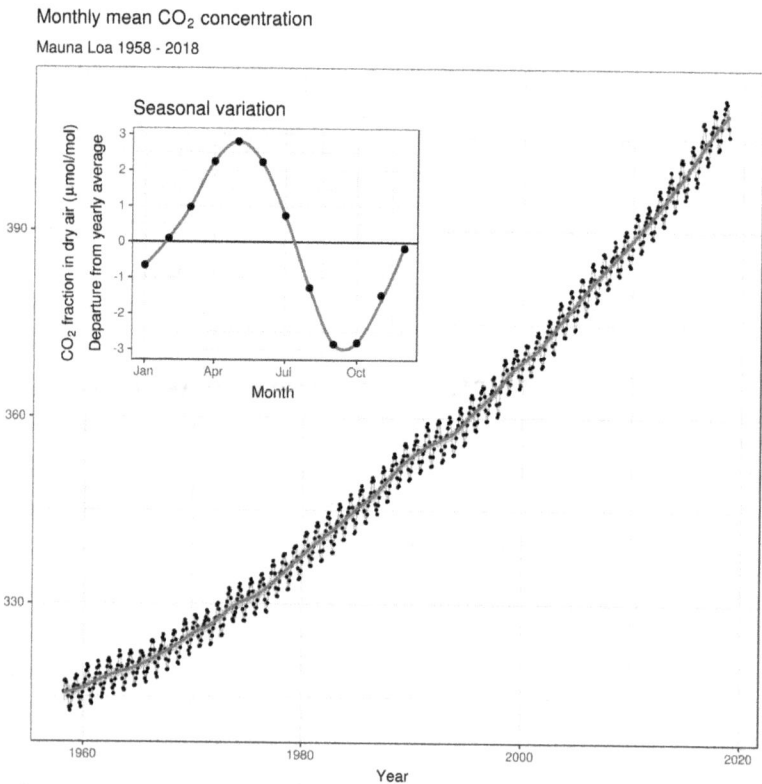

The Keeling Curve showing the rise in carbon dioxide levels from 1958-2018.
Courtesy Dr. Peter Tans, NOAA/ESRL

We know the source of the extra carbon dioxide levels is from the burning of fossil fuels. There are no observations that the rapid rise in carbon dioxide levels comes from natural sources, including an increase in the rate of volcanism. There are several ways we can determine the source of the extra carbon dioxide and rule out natural sources.

First, all elements come in slightly different varieties called isotopes that vary in the number of neutrons in the nucleus. Molecules of carbon dioxide from volcanic sources have different ratios of isotopes compared to coal and oil. When scientists measure the carbon dioxide in the atmosphere, it has the tell-tale signature of coming from coal and oil.

Secondly, we see no more volcanic activity compared to a hundred years ago to explain the rise of carbon dioxide in the atmosphere. Also, we put about 2.2 million pounds of carbon dioxide into the atmosphere every second of every day, which translates to approximately 35 billion tons of carbon dioxide each year. Based on these numbers, it becomes clear where the extra carbon dioxide comes from.

Climatologists have also ruled out other natural factors of climate change. For example, the continents have moved less than 10 feet since the beginning of the industrial revolution, not enough to affect ocean currents or other climatic patterns. Those kinds of changes require millions of years. Solar output does vary over time; in fact, solar output entered a period of low activity, which should be causing lower temperatures. Despite the lower solar activity, we have actually seen an increase in surface temperatures. And lastly, based on the Earth's orbit, we should remain stable for a few thousand years more, or perhaps become slightly cooler

The science behind climate change follows the same methodology as any other branch of science. We make observations about the world, formulate hypotheses to explain the observations and then go and test our hypotheses. In the case of climate change, we have observed a warming climate. All the natural sources of climate change (solar activity, continental movement, increased volcanism, and Milankovitch cycles) have been disproved by scientific observations and experiments as a cause of the current warming trends. The only valid scientific explanation remaining to explain the rise in temperatures is the burning of fossil fuels, and deforestation. The burning of fossil fuels and clear-cutting are raising carbon dioxide levels in the atmosphere, driving the rise in surface temperatures.

Accurately Predicting Climate Change Can be Difficult

All the variables affecting climate change make accurate predictions difficult. For example, the climate seems to be susceptible to positive feedbacks. The Snowball Earth was probably the result of positive feedback leading to global glaciation. For a modern example of a positive feedback, I will use the melting of the Arctic sea ice. Being white, ice reflects much of the sun's energy back into space keeping it cooler. As the ice melts, it leads to more open water, which is much darker, absorbing more energy, causing the water to warm. The warmer water, in turn, melts more sea ice in the summer and slows ice formation in the winter. In this case, the melting of the sea ice amplifies the rate at which it melts in a positive feedback. Additionally, the loss of Arctic sea ice makes the region much warmer, which in turn melts the permafrost in the tundra releasing more carbon dioxide into the atmosphere, once again increasing the warming in the region. Also, warmer water holds fewer gases, so as the oceans warm, less carbon dioxide is absorbed by the oceans. These positive feedbacks can quickly lead to rapid warming.

Another example of the difficulties in predicting climate change lies in something as seemingly simple as clouds and cloud formation. Water evaporates faster in warmer climates, in turn, more water is in the atmosphere leading to more clouds. However, clouds are white and reflect sunlight into space, but at night they hold heat from the day. That's why cloudy days and nights have less temperature swings. However, not all clouds are created equal, it turns out that the type of cloud, its thickness, and height are important in determining whether or not the increased cloud cover reflects more sunlight or holds more heat. Understanding the relationship between clouds and warmer climates remains an active area of climate research.

Climatologists must balance sources and sinks of carbon dioxide. The rate of photosynthesis, a sink, has to be compared to the amount of fossil fuels being burned each year. They must take into account deforestation that adds carbon dioxide and makes the clear-cut areas warmer. Yet another factor to consider is that as the permafrost melts, the northern boreal forests moves northward. As primary productivity increases, more carbon dioxide is removed from the atmosphere, but the forests are darker and absorb more sunlight than the tundra, so the region warms more. As the permafrost warms, the peat will be broken down, releasing more carbon dioxide.

Warmer temperatures also affect global circulation patterns that distribute incoming solar energy, strongly influencing regional climates around the planet. For example, a warmer Arctic in the winter can bring very cold air to the northern hemisphere. A polar vortex sets up as the Arctic cools in the winter, the vortex prevents much of the cool air from spilling southward. However, temperatures that are 20°F warmer in the Arctic winter weaken the polar vortex, it allows the Arctic air to spill southward. Although, the Arctic air may be above average in warmth, the air is still cold compared more southern latitudes of Europe and North America.

Climatologists look to the past to reconstruct the climate by using proxy data similar to what was used to create the Hockey Stick Model to help make predictions about current climate change and sea level rise. Taken together, these independent sources of information paint a picture of the Earth's climate through time. Geological and fossil records are useful for constructing ancient shorelines and the types of animals that were present, or went extinct. For example, when the Earth was much warmer, Antarctica was forested and ancestors to alligators lived as far north as Wyoming.

Based on the historical proxy data, we know for the last 850,000 years, there have been cycles of maximum glaciation with periods of shorter interglacials. Times of glacial maximums coincide with sea levels nearly 300 feet lower than today, in contrast times only 3.9°F warmer coincided with sea levels 20 feet higher. In fact, the last time carbon dioxide levels were similar as today, the Earth was only 3.9°F (2°C) warmer and the sea levels were about 20 feet higher. These observations about rising sea levels and temperatures worry scientists about the rapid rise in carbon dioxide levels observed today. We have already warmed the Earth 2.4°F

(1.3°C) over the last century, with much of that warming taking place in the last 30 years. And to make the point again, this is the fastest changes observed in the Earth's climate, it's ten times faster than the end of the last ice age.

In the past 20 years, our knowledge of climate change has rapidly improved, along with that our ability to make predictions about its future. While there is often disagreement among scientists about which factors are most important in predicting climate change, they all agree that climate change is here, it is not just a future problem. While uncertainties exist in our ability to predict the future, we know that past patterns in the climate clearly show that sea levels will rise and rainfall patterns will shift. We just don't know how fast the climate will warm, sea levels will rise, or how much rainfall patterns will shift. However, one trend in climate predictions stands out as scientists continually revise their models, each time, they underestimated the rate of climate change. That means the rate of climate change and its effects are rapidly rising, and they are happening faster than scientist thought a decade ago.

The Implications of Climate Change

What are the implications for climate change to life on the Earth? To a first approximation, rapid climate change caused the most severe mass extinctions. The End-Permian Extinction 251 million years ago, brought about by a period of intense volcanism followed by rapid climate change, was the greatest extinction event in the Earth's history. The oceans may be particularly susceptible to climate change. When water conditions change, many animals cannot adapt rapidly enough and perish. Intense volcanism causes rapid changes to the water chemistry. For example, higher levels of carbon dioxide in the atmosphere lowers the pH of the oceans making them more acidic. The lower pH makes it hard to extract calcium carbonate from the water to form shells or reefs,

Today, the oceans are acidifying as the pH drops as carbon dioxide dissolves into the water creating carbonic acid. In fact, it's been estimated that up to half of the carbon dioxide released in the last hundred years has been absorbed by the oceans. The oceans cannot continually absorb carbon dioxide from the atmosphere at the same rate they have for the last century. Once the rate carbon gets absorbed into the oceans slows down, atmospheric carbon dioxide levels will rise even more rapidly today, causing even faster rates of climate change.

In the tropics, coral reefs grow in crystal clear waters that rarely experience large temperature changes. The corals live in these nutrient poor waters because they form a symbiotic relationship with photosynthetic zooxanthellae that live in their tissues. When corals become stressed, they often lose their symbionts and take on a white appearance, hence the name coral bleaching. Higher sea surface temperatures cause coral bleaching, and in severe cases, leads to death of the corals. In 2015, NOAA declared the third ever global coral bleaching event due in large part to the oceans warming and one of the largest El Nino events on record. When you consider that coral reefs are home to nearly 50% of all marine species, the implication for the loss of diversity becomes apparent.

Tropical reefs are not the only marine ecosystems being affected by climate change and ocean acidification. At the poles, warmer water not only decreases the ice cover, but it has increased the productivity of phytoplankton by an estimated 40-50% due to the longer growing season. In some ways, this is good news because it will help remove carbon dioxide from the atmosphere and could serve to boost fish populations in the region. Many marine fish easily move into new regions as climate change results in warmer waters poleward. As the Arctic and Antarctic waters continue to warm, new species of fish move in at the expense of native fishes and other marine organisms. The impact of these changes in fish populations will not be well understood for some time.

Climate change is more than just a warming of the atmosphere. It is also a shift the patterns of weather. Some places will become wetter while others will become drier. Entire ecosystems could potentially be lost or shift in location. For example, warmer temperatures are leading to longer growing seasons and a melting of the permafrost, allowing trees to grow further northward at an unprecedented rate, shrinking the Arctic tundra. To complicate matters, the northern boreal forest may be expanding northward, but its southern limits may be reduced by increased forest fires resulting from warmer temperatures.

In the western United States, the snow pack is expected to decrease by 50% over the next 50 years. The reason for the loss of the snow pack is not necessarily a loss of precipitation, but warmer temperatures that will melt the snow earlier in the season. Rapid spring melting of the snow pack could cause several problems for the region. For example, less water will seep into the ground, potentially stressing the trees in the summer as temperatures warm and water is lost through transpiration. Second, mountain streams normally have higher water flow in the spring from snow melt, but maintain constant flows throughout the summer because snow melt in the mountains recharges the groundwater. The loss of winter snow pack could significantly alter the flows of western rivers where the rapid spring snow melts would cause very high flows followed by very low flows, or even drying of the river bed in the late summer.

On land, some species of animals will be able to track the changes in temperature and adjust their distribution accordingly. This is the climatic envelope hypothesis, which basically means that a species is restricted to a particular climate, similar to their fundamental niche. When the climate changes, their fundamental niche changes and their populations change in response to the new climate. We have already seen changes in species distribution for marine fish, butterflies, birds, and plants as they track their climate envelope.

However, not every species will move to new regions. One good example are cold water trout who cannot survive temperatures much above 77°F (25°C). As temperatures warm in the summer above this critical temperature, these fishes may be unable to migrate to cooler waters and will perish. A criticism of the climate envelope hypothesis is that it only considers a single species. However, a recurring theme of life is that no species is an island, isolated from other species. If a predator is dependent on a particular prey item and that prey cannot track its climate envelope, then the predator must adapt to a new food source, or perish. In another scenario, imagine a flower that has evolved with a specific pollinator; if the flower cannot survive under a new climate, nor track its climate envelope, it will perish taking the pollinator with it.

The effects of climate change will not be limited to plants and animals, it will severely impact humans and our civilization. Changes in rainfall patterns will disrupt our crop production leading to food shortages. Warmer temperatures cause drought and cities in desert regions will face water shortages which will certainly cause problems. As sea levels rise, coastal cities will face more flooding causing a loss of economic value for homes in businesses. Places like Miami Florida with millions of people will eventually be flooded, and you can't simply build a sea wall for protection. Not only will people lose money from the inevitable economic collapse, but they will be forced to move. Climate refugees will be forced to move from areas that were flooded or became too dry. Unfortunately, the mass migration of people leads to increased nationalism and more conflict, a problem already happening today.

In summary, the world is already responding to current climate change. Many species will simply move to suitable habitats while others will perish, becoming extinct. Entire ecosystems will be altered as new species move in and others are lost. Based on what has happened to life during previous episodes of climate change, we will experience another, or 6th mass extinction event in the next few hundred years. The one thing that separates this climate change event and the ensuing mass extinction is the presence of humans and how we greatly altered the Earth, the topic of the final chapter.

Chapter 14

The Human Impacts to Life: How we are changing the world

Destroying rain forest for economic gain
Is like burning a renaissance painting to cook a meal
E.O. Wilson

The last word in ignorance is the man who says
of an animal or plant,
What good is it?
Aldo Leopold

Introduction

For nearly 4 billion years, life has existed, evolving, diversifying, and changing the Earth. Over time, new species evolve to take advantage of new environments, while others go extinct because they were out-competed or cannot adapt to the new environment. Mass extinctions caused by climate change or meteor impacts have occurred at least 5 times in the past. After each event, life once again diversified to fill the empty niches. Every new species can potentially alter the environment, and so it is with a special bipedal ape with an unusually large brain and an uncanny ability to alter the environment.

Conservation biology is the scientific study of biodiversity with the goal of protecting species and the ecosystems they reside in. It's a difficult field for several reasons. First, understanding how all the different species interact with each other in the ecosystem can be quite difficult and may take years to understand. Second, scientific studies often become quickly politicized because of their economic and cultural impacts. And lastly, not everyone places the same value on wilderness areas and diversity; unfortunately, this view is largely out of ignorance. As our world has become increasingly urbanized there has been a growing disconnect with nature as we preoccupy ourselves with our daily lives, social media, and content-free news.

Human activities today are causing the 6th mass extinction. Although it may be decades before we fully realize the extent of this extinction, for it is just starting to get underway. I begin this chapter exploring the causes and consequences of the previous mass extinctions. Most of this chapter is a brief introduction explaining how our activities are degrading ecosystems leading to extinctions. The last section is my opinion on the need for conservation, the future of humanity, and life on the Earth.

A Brief History of Mass Extinctions

The Earth may be 4.6 billion years old, but the last 0.6 billion years has witnessed the rapid rise of complex life and increasing diversity. It may have taken 4 billion years, but once animal life appeared in the oceans, it rapidly diversified into cnidarians, mollusks, arthropods, echinoderms and chordates, along with a lot of different types of worm-like animals. Once plants, fungus, and animals colonized the land, diversity continued to increase, in turn, creating more complex ecosystems. However, over the past 542 million years there have been some pitfalls to the slow, but continual rise in diversity.

The first well known mass extinction occurred at the end of the Ordovician Period 444 million years ago as the world experienced an intense ice age. The cooler temperatures led to extensive glaciers that lowered sea levels, which in turn reduced shallow water habitats, causing the extinction of many marine organisms, especially the mollusks. But, life rebounded over the next 70 million years and diversified as plants colonized the land eventually evolving to grow tall and forming the first forests. Finally, the land became green and animals followed the plants, and they too diversified.

The Earth's second mass extinction about 365 million years ago marked the end of the Devonian Period. So far, the reasons for this mass extinction remain unclear. Although some evidence suggests that a period of intense volcanism warmed the climate and acidified the oceans. At the same time, several continents converged together allowing species that were once isolated to mix for the first time. As a result, the spread of better adapted cosmopolitan species may have been a factor in this second mass extinction.

A similar, but smaller extinction took place when South and North America were joined for the first time about 5 million years ago when the Isthmus of Panama formed. Many North American species migrated southward driving the extinction of South American species that had been isolated for millions of years.

The End Permian extinction 251 million years ago marked the third and largest mass extinction in the history of the Earth when an estimated 85-95% of all species went extinct in a time period of a few hundred thousand years. It brought about the end of the Paleozoic Era and ushered in the Mesozoic Era. This extinction was likely caused by a period of intense volcanism creating the Siberian traps, which holds 720,000 cubic miles of volcanic rock. That would be enough to cover 23% of the lower 48 in rock one mile deep! The intense volcanism drastically changed the climate and acidified the oceans. It took about 20 million years for diversity to fully recover from this mass extinction. Ironically, prior to the extinction, the ancestors to modern mammals dominated the land. But after the extinction, the reptiles evolved into the dinosaurs and dominated the land.

The Mesozoic Era suffered the fourth mass extinction about 201 million years ago possibly due to another period of intense volcanism. This mass extinction didn't wipe out the dinosaurs, but it did cause significant losses to marine life. The fifth and last mass extinction occurred about 65.5 million years ago and was the second worse extinction in Earth's history. This one is perhaps the most famous for wiping out the dinosaurs after a successful 170-million-year reign. For several decades, the leading hypothesis to explain the loss of the dinosaurs has been a large meteor impact that devastated the Earth's ecosystems. However, a few million years prior to the meteor impact, the dinosaurs were rapidly declining, quite possibly due to another period of intense volcanism. This time, in a region of modern-day India, known as the Deccan traps. The extinction of the dinosaurs may have been caused by this one-two punch from rapid climate change brought about by volcanism and a meteor impact.

From these "Big Five" mass extinctions, there are several take home points. First, rapid climate change was almost certainly a cause for several of them, including the most severe. Second, diversity does recover after mass extinctions, but it takes millions of years to do so. Lastly, previous mass extinctions have paved the way for the evolution of new species millions of years later, such as the rise of the mammals after the extinction of the dinosaurs.

The 6th Mass Extinction

Today, we are causing the 6th mass extinction. Throughout time, there has always been a background extinction rate, which is about 1 species per 100 years per 10,000 species. That means if you live in an area with 10,000 species, about one would go extinct every hundred years, but the loss would be offset by the evolution of new species. The world today has an estimated 8.7 million species, so we would expect about 8-9 species to go extinct worldwide each year with natural extinction rate. In just the last 150 years, three times as many birds and mammals have gone extinct than in the previous 200 years.

However, the current extinction rate is closer to 50 species per 100 years per 10,000 species. That means we are losing an estimated 400 species each year to extinction. The loss of species of plants and animals is so fast that scientists are unable to fully describe them before they go extinct, let alone learn anything about their lives or potential benefits. Some conservation biologists estimate that we will lose 65% of all species by 2100, or 5.6 million species in 100 years, which is about 56,000 species per year! The extinction rate has probably not reached such a high rate yet, but it is not an impossibility in the near future based on the impacts humans are having on the planet.

It can be hard to know when a species is extinct, many are rare or elusive, or just plain hard to find. Most scientists do not like to declare a species extinct, even if they have not been able to find it after a great deal of effort. Nowhere is this more apparent than the amphibian declines in the last few decades in Central and South America. A fungus combined with habitat loss, pollution, and changing climates has caused severe population declines in many species of amphibians. Estimates based on their current global population declines indicate that their current extinction rate is 25,000 times higher than the natural background extinction rate for amphibians.

Other vertebrates are also suffering losses, one out five species of fish are at risk of extinction and nearly 50% of all mammal species are declining. About 14% of all birds are at risk, but birds may have already suffered through a major round of extinction about 2,000 years ago. By bringing livestock and rats, combined with over exploitation, it's been estimated that the spread of people across the Pacific caused the extinction of nearly 2,000 endemic bird species on isolated Pacific islands.

We stand to lose a lot, so the question becomes, HOW? How are humans causing one of the largest mass extinctions in the history of the planet? Unfortunately, there are lots of reasons how we are causing it, but it comes down to a few critical problems that the world will have to eventually address. The biggest problem is the rapid population explosion of humans based on cheap energy from fossil fuels. Human populations are rapidly expanding and using resources at an unsustainable rate leading to rapid climate change, habitat loss, a spread of invasive species, overexploitation of resources, and pollution. And in many cases, these factors are confounding, further exacerbating population declines of wildlife and extinction rates.

The Human Population Explosion

To say that humans are like no other animal the world has ever seen could still be an understatement. Every species is unique; representing an unbroken lineage going back to the origin of life. We are a mammal and evolved by natural selection just like every other animal on the planet. But, humans are unique in several key ways. Unlike all other animals that have ever preceded us, we are the smartest in that we can alter our surroundings to our liking. We no longer have to evolve to match our environment.

All other species must reside within their fundamental niche and most are further restricted to their realized niche by competition or other biological factors. This is not the case for humans, if it's too cold for us, we put on warmer clothes, build a fire, turn up the heater, or drive to southern Florida in a giant RV for the winter. Likewise, if it becomes too hot, we wear lighter clothing, move indoors, turn on the AC, or drive back north in a giant RV for the summer. Humans are not restricted to living in a single habitat, in fact we live on all continents, and in every ecosystem. We are not dependent on a single source of food, but can eat just about anything.

Since the end of the last ice age over 10,000 years ago, our population has continually grown and spread around the planet. Much of the growth can be attributed to a period of stable climate and the agricultural revolution, which freed people from the vagaries of unpredictable climates and a hunting and gathering lifestyle. It offered people more stability and opportunities to take up other professions. From this, civilization was born. It took nearly 200,000 years of human existence for the population to reach 1 billion in 1804, 123 years later it doubled again to 2 billion in 1927. The baby boomers born after World War II between 1945 and 1964 were the first generation of people born to experience a doubling of the population in a mere 33 years between 1960 and 1999, when the population reached 6 billion people. As of 2018, the world population has reached 7.7 billion people and continues to grow at a rapid rate of nearly 80-90 million people per year. Fortunately, there are signs that the world's population growth rate is beginning to slow.

Two factors in combination are largely responsible for the rapid rise in human populations; a cheap source of energy found in coal and oil and the technology to use it. Although humans are unique in our intellectual abilities to exploit or even create novel resources, nevertheless human populations have responded to a rapid increase in available energy just like any other animal. The ability to cheaply grow and distribute food because of cheap fossil fuels temporarily expanded the Earth's carrying capacity for humans. Today, we are burning through fossil fuels at an alarmingly rapid rate, returning carbon dioxide to the atmosphere that has been sequestered for hundreds of millions of years.

Our amazing rise as successful species has come at a price that we will pay. Eventually, we will run out of fossil fuels and unless we replace it with another cheap energy source, our population will decline to match the carrying capacity of our Earth, there is no way around this. Secondly, we are drastically altering our climate with devastating consequences, and lastly we are having a devastating impact on the world's diversity as we cut down forest to grow food and make room for our expanding population. Through our activities, we are directly causing a mass extinction, and unlike any other mass extinction, this is the first one caused directly by the actions and success of a single species.

Global Climate Change

The global climate is changing, rapidly, due almost entirely to the burning of fossil fuels returning carbon dioxide to atmosphere that has been sequestered for hundreds of millions of years. Climate change is a scientific fact backed by repeated observations and experimental verification. Carbon dioxide levels have been rapidly rising in the atmosphere from the burning of fossil fuels and there is no other explanation to account for this change in the last century.

Climatologists have reconstructed the Earth's climate going back thousands of years and in some cases, hundreds of thousands of years. Regardless of the data used, whether it's from ice cores, tree rings, lake and marine sediments, or direct measurements, it all paints the same picture: ***The climate is rapidly warming.***

Rapid climate change has been implicated in some of the largest mass extinctions of the Phanerozoic Eon (542 Million years ago – present), including the End Permian Extinction that wiped out 80-95% of all life on the Earth. Species go extinct during times of rapid change because they cannot evolve quickly enough to survive in the new environment or they are unable to migrate to new regions that are suitable to their physiological tolerances. For example, if you require wet and humid environment and it quickly dries from loss of precipitation; then you must evolve to survive in the drier environment, or migrate to wetter regions. If a species fails to adapt or move to suitable habitat, it will go extinct.

Today, there is an estimated 8 million species, give or take a million. Some will be able to migrate to new habitats as the climate alters their habitats, but many will not. Additionally, the rate of climate change is so fast that many species will not be able to evolve to survive in their current locations and will face extinction. Global climate change may be one of the biggest factors leading to the loss of diversity worldwide.

Habitat Loss and Degradation

Imagine a priceless painting, *Starry Night* by Van Gogh. Intact it's priceless. Now imagine you cut it into small pieces. Every time a part is cut and removed the value of the painting is reduced, the small pieces by themselves are practically worthless. The same goes for ecosystems; intact they are priceless, as they become fragmented with roads, urbanization, or clear-cutting, their value becomes increasingly diminished. Losing species to extinction is like removing the brush strokes, at first, it's not very noticeable, but with every loss, the quality of the picture loses its meaning and worth. As ecosystems lose species, they become a shell of their former selves.

Forests cover about 31% of the land on our planet. Each year, about 46-58 thousand square miles are lost to clear cutting, logging, fires, and degradation from 225, loss of disturbance, or climate change. That's an area about the size of Alabama lost, much of that occurs in the tropics where species diversity is the highest. Clearing the Amazon rain forest in Brazil is one of the biggest areas of concern. The land is being cleared mostly for cattle ranching, and small-scale clearing for subsistence agriculture. When you realize that tree diversity can reach 300 species per square acre in some regions, it's easy to see how the loss of forests are driving the loss of species. There are many more species of animals compared to plants. To estimate insect diversity in the tropics, an entomologist fumigated a single tree in Panama and collected 50 new species of beetles that were previously unknown to science.

Road construction fragments the landscape and alters the dynamics of the forests and making it inhospitable for many species. Studies of forests patches in Panama and in the Amazon have shown that smaller patches of forest house fewer species over time. What this means is that 10 forest patches, 10 acres each will have fewer species than a single large patch of 100 acres. Some animals, including large predators, require large intact tracts of land to survive and aren't adapted to a fragmented landscape. You would think that birds could easily move between forest fragments to maintain their populations, but some birds won't fly short distances across fragmented habitats.

The North American grasslands once formed vast prairies with herds of buffalo grazing on the native grasses. Plant diversity remained high from frequent fires and grazing that prevented any one species from becoming dominant. Today, the large buffalo herds are long gone, hunted to the brink of extinction and the land has been converted to croplands. In Texas, Oklahoma, and New Mexico improved technologies in oil and natural gas extraction led to a rapid rise in oil pads throughout the region. As you can expect, plant and animal diversity, especially grassland birds, such as prairie chickens and Sage Grouse, have declined rapidly in the past few decades from the loss of habitat from farming and resource extraction.

Habitat loss isn't limited to clear-cutting and habitat fragmentation. The loss of natural cycles can also degrade habitats leading to a loss of diversity. In the southeastern U.S. routine fire would spread through Longleaf pine habitats about once every 2-3 years. Fire was an integral part of these ecosystems, a single fire started from summertime lightning strikes could last for weeks. As the fire burned across the landscape it removed undergrowth and fallen logs preventing the buildup of fuel on the ground. The frequent fires maintained an open park-like forests promoting high rates of plant and animal diversity. In some locations, plant diversity reached a high of 50 species per square meter.

Beginning in the 1920s, active fire suppression resulted in the encroachment of hardwoods, which further prevented the routine fires and reduced diversity. The loss of fire also coincided with increased habitat fragmentation, and as a result, the Longleaf pine forests of the southeast have been severely degraded over time either through clear cutting or succession to hardwoods. Species like the Red-cockaded Woodpecker became endangered because they are totally dependent on pristine Longleaf pine ecosystems with natural fire cycles and an open understory. These small woodpeckers are not the only ones that have suffered losses from ecosystem degradation, other species including Indigo Snakes and Gopher Tortoises also dependent on the pine habitat have been severely reduced in populations. Today, the Longleaf pine ecosystem is one of the most endangered ecosystems on the planet.

Habitat loss and degradation are not limited to the land. Our streams and rivers have also become severely impacted by dams and the lakes they create. Most rivers naturally vary in flow throughout the year; a type of natural disturbance creating habitats in rivers and maintain connectivity to riparian areas. Throughout much of North America, high flows often occur predictably in the spring from snow melt. High flows can also occur less predictably when rivers swell due to increased rainfall.

It is this natural flow variability, both predictable and random, that is important for maintaining riparian areas adjacent to the rivers and creating different types of habitats in the river itself. Dams fragment streams, reduce flow variability, confining rivers to their main channel and separating them from their adjacent riparian areas. The loss of spring flooding is similar to the loss of fire Longleaf pine forests, the change in disturbance regimes simplifies the ecosystems and results in a loss of diversity.

Many species of fish including trout, sturgeon, and eels rely on rivers being connected to the oceans so they can complete their life cycle. Eels and sturgeon must return to the oceans to reproduce, whereas salmon must return to rivers to reproduce. If dams impede their migrations, then those migratory species will become extirpated from those rivers. Additionally, some species of fish produce semi-buoyant eggs that must float down a river to suitable habitat where the fish hatch, grow, and then swim back upstream. Large dams produce lakes that are unsuitable to these pelagic spawning fish. They lay their eggs, but they float down into lakes where the young fish are unable to survive.

Perhaps one of the most devastating forms of habitat destruction that goes largely unseen is trawling on the continental shelf, especially for shrimp. Trawling involves dragging nets across the bottom of the ocean behind a large boat. These nets indiscriminately catch everything in their path, while destroying the marine habitat in the process. In 2001, satellite trackers were used to monitor and map the movement of boats trawling for shrimp in coastal waters. It was discovered that they trawl an area about the size of the entire continental shelf every 18 months. In some places it takes decades for the area to fully recover from a single trawling incidence. Unfortunately, the habitat loss goes largely unnoticed by everyone because it is underwater and not readily observed unlike deforestation that can easily be seen or tracked with satellite imagery.

Invasive Species

When a species is introduced into a new habitat, becomes established and spreads to new areas while displacing native species, then it is called an invasive species. When a species becomes invasive, it displaces the native species disrupting entire ecosystems. Common examples of invasive species you encounter almost every day includes pigeons, House (English) Sparrows, and European Starlings, all brought over from Europe.

There are several reasons why a species becomes invasive. Often, invasive species are habitat generalist and take advantage of degraded ecosystems. Sometimes it's because they escaped their natural enemies allowing their populations to grow and expand unchecked. Or the invasive species is a superior competitor compared to the native species.

In the Eastern United States, the aquatic plant *Hydrilla* spread rapidly in both lakes and rivers, displacing the native vegetation and altering fish communities. Its thick growth limits fish populations and skews them to smaller individuals. Bird populations dependent on native fish and snails can also be harmed by the presence of *Hydrilla*, as it completely degrades the habitat for most native organisms.

The deciduous forests of Eastern North America have suffered through several major tree die-offs in the last hundred years. In the early 1900s, a pathogenic fungus spread rapidly throughout the east wiping out native American Chestnut trees by the 1940s. Prior to the chestnut blight, the American Chestnut tree was the dominant hardwood of the region. More recently, the woolly adelgid, a small insect introduced to Virginia in the 1950s, has decimated the Eastern Hemlock tree. It feeds by sucking the sap of hemlock and spruce trees, eventually killing the tree. As of 2015, 90% of the geographic range of the hemlock trees is threatened by the woolly adelgid. Its spread northward has been limited by extreme cold winters, however recent warming from climate change may allow this invasive species to move northward, killing more trees.

Remote islands in the Pacific are home to endemic birds found nowhere else in the world, and they also serve as nesting grounds for millions of sea birds. Having never been connected to mainland areas, these islands were free of many predators including rats and snakes. During World War II, cargo and war ships fighting in the Pacific accidentally transported brown tree snakes to the Island of Guam. Within a few years the snake's population exploded due to lack of natural predators and abundant resources. The result of the introduction was that they wiped out almost all the endemic birds on Guam in a few decades.

Introduced rats are a major problem for nesting birds on oceanic islands. Mammals don't naturally colonize remote islands because they cannot survive drifting for days on the ocean, they die after a few days without food or water. Unfortunately, humans have spread rats to oceanic islands where they have become established causing great harm to the native plants and animals. Rats are omnivores, a food generalist capable of eating a variety of foods including plant and animal material. They cause harm to plants by eating their seeds and fruits. Native bird populations on islands evolved in the absence of natural predators, so they are mostly defenseless against rats that eat their eggs. For example, most seabirds nest on the ground, where the rats are known to eat the eggs and the young. To make matters worse, rats easily climb trees so they can eat bird eggs and nestlings in the trees and on the ground.

The eradication of invasive species can also become quite complicated and actually cause harm to the remaining native species. When invasive plants displace native plants, some native animals will form new relationships with the invasive species. In the desert southwest, the Willow Flycatcher successfully nests in non-native salt cedars that have replaced much of the native vegetation in riparian areas. If managers attempt to remove the salt cedar too fast without replacing it with other vegetation, then its removal could negatively impact the Willow Flycatcher. To further complicate the problem, the loss of natural flows and prolonged drought has made conditions favorable for the salt cedar and less favorable for native vegetation, including cottonwoods

Overexploitation

Standing at the ocean's edge, its vast expanses seem to go on forever. To some, it would seem almost impossible to deplete its resources, yet that is exactly what we are doing. The detrimental changes taking place under the waves remain largely unseen and are due largely to overexploitation of its resources

Overfishing is a huge problem worldwide and is causing major changes to the marine ecosystems. Unfortunately, the issues causing overexploitation are complicated and not easily solved. Part of the reason for overfishing arises from the fact that fish are a great source of protein. In fact, salmon has been labeled a super food, or one of the healthiest foods in the world, so we should be eating more of it.

Because fish are a highly desirable food, economics becomes a major factor driving declines in fisheries. From a business point of view, if you have one fishing boat and turn a good profit, then it's in your economic interest to reinvest the money into more fishing boats to make more profits. Eventually, the fishing fleets grow larger, taking more and more fish from the oceans.

Another problem causing the decline in fisheries from what is called the tragedy of the commons. If there is a common resource that is unregulated, then it is in your best interest to fully exploit the resource, or someone else will exploit it. Fish in the open ocean are a common resource available to everyone. Unfortunately, they are overexploited because the biggest immediate benefit goes to the one who most exploits the resource. The drive for short-term profits come at a cost of long-term sustainability.

Additional problems arise because of the difficulties in obtaining accurate information on annual takes of a commercial fish, or to determine the size of their populations. Large-scale industrial fishing can collapse a fishery within ten years to one-tenth its original size. Most commercial fish populations can sustain a small harvest, but when too many fish are taken, their populations quickly collapse as in the case sardines, anchovies, and the Atlantic Cod. In 1992, populations of Atlantic Cod dropped to 1% of their historical levels, driven mostly by improved equipment and technology.

The collapses in fish populations are especially troubling in long-lived species including sharks and other large apex predators. Studies have shown that more than 90% of larger fish have been removed from the oceans mostly due to long-lines, drift-nets, and other industrial fishing practices. Large tuna and sharks have been hit especially hard. Shark populations are collapsing world-wide, and some species have declined 97-99%. Perhaps the saddest part of the shark story is that most of them are caught only to have their fins removed to make shark-fin soup. The rest of the animal is thrown overboard to die.

The impact from the loss of large apex predators on marine ecosystems is just now being fully understood. For example, parrotfish are found on coral reefs throughout the world where some species eat parts of the coral, creating sand and preventing algae growth. One of their natural predators are sharks. Unfortunately, the decline in shark populations has led to higher populations of parrotfish, which in turn, eat more corals. As you can see, the loss of sharks leads to further degradation of coral reefs by allowing one species to become overabundant, directly causing harm to the reef.

Trawling for shrimp, a problem I discussed in habitat degradation, is also a problem of overexploitation. Not only does trawling destroy the habitat of the target species, including shrimp, it also catches every species in the area. Marine sea turtles are especially hard hit from shrimp trawling as they get caught in the nets and drown. Sea turtles also face problems on their nesting grounds as local people often collect and eat the eggs, further exacerbating their population declines. As a result, all sea turtles in the US waters are listed as endangered species.

On land, overexploitation is also a problem. In the 1800s, the Passenger Pigeon in North America may have been one of the most abundant birds in the history of the world with population in somewhere between 3.5 and 5 billion birds. They were wiped out by widespread commercial hunting and habitat loss. Between 1870 and 1890 their populations went into rapid decline and by the 1890s their populations had declined to the point where extinction was all but a certainty. The last Passenger Pigeon died in captivity in 1914. The extinction of the Passenger Pigeon serves as hard lesson that even very abundant species can be driven to extinction through rampant overexploitation and loss of habitat.

Along the Atlantic and Gulf of Mexico shores, horseshoe crabs crawl out of the water during spring high tides to lay their eggs at the water line as they have done for 450 million years. Thousands of migrating shorebirds time their migration to coincide with the laying of the horseshoe crab eggs. With the collapse of the lobster industry (from over harvesting), fishermen have turned to eel fishing and collecting horseshoe crabs for bait. At its height in the late 1990s, over 2.5 million horseshoe crabs were collected. As their populations rapidly declined, so did the shorebirds that depended on them, providing another example showing that species do not exist in isolation. Species form communities where the loss of one species can directly lead to the loss of other species. In this case, the Red Knot populations declined by over 70% in the last decade due to the loss of horseshoe crabs from over-harvesting.

Over-harvesting does not stop with catching animals for food or bait. Collectors for the pet trade impact reptiles, including box turtles, snakes, and lizards. To make matters worse, collecting reptiles is not well regulated, and there's little data or reporting on harvesting for the pet trade, so we don't know how many animals are being removed each year.

Like many people, I enjoy watching colorful marine fish in my tanks. But, the aquarium trade presents another problem where collectors harvest tropical marine fish from the wild. Some areas such as Australia and Hawaii closely regulate the taking of tropical fish, attempting to keep the industry sustainable. Unfortunately, it's not so well regulated in other regions where local people use dynamite or cyanide to "stun" the fish. The problems with exploding dynamite underwater are apparent when fish and corals are immediately killed. The problems with cyanide are more subtle, most of the fish harvested with cyanide usually die within a few months from liver failure and reefs exposed to cyanide will begin to die. Both methods of fish harvest are unsustainable for the simple fact that the collecting methods destroy the habitat required for the fish they are harvesting.

Over-harvesting is not limited to animals in the pet trade. Currently, populations of many cactus species in the southwest and Mexico are in sharp decline from unregulated collecting. Unregulated collection of cacti and other rare plants can decimate local populations in a single day. The cacti are sold to collectors mainly in the U.S. and Europe. Similar to cacti, orchids face similar problems of over-harvesting in the wild

The problems do not stop here; bears, rhinos, and sea horses are collected for traditional Asian medicines. In African countries suffering from war, famine, or poverty, people turn to the local mammals and hunt them for "bush meat", causing the bush meat crises where many large primates are being killed for food. Most recently, Elephants declined by 30% between 2007 and 2014 from illegal poaching for their tusks to be sold in Asian markets. The problem has become so bad, that elephants are evolving to have smaller tusks.

Pollution

We have become a disposable planet. The waste and byproducts of our civilization is staggering. The widespread use of plastics has become a major problem because they are cheap to make, but they take a long time to break down. Each year, about eight million tons or 16 billion pounds of plastic are dumped into the oceans each year. The plastic doesn't go away; instead plastic waste gets consumed by animals causing it to accumulate in the food chain. So far, we have found at least 700 species of marine animals have eaten or ingested plastic. Plastic just doesn't make animals sick, it can kill them too. Tragically, plastic bags resembling jellyfish get eaten by giant Leatherback sea turtles. Seabirds and whales are killed by plastic debris when they accidentally eat it. They aren't equipped to handle the plastic and end up choking to death on it.

Pollution from coal-fired plants releases nitrogen and sulfur compounds into the atmosphere where they react with water to form acid rain. In the eastern U.S., acid rain from coal-fired plants and automobiles has caused fish kills in some lakes and has acidified the ground so much in some places it has even killed trees. Coal also contains heavy metals including mercury, lead, arsenic, and cadmium. Modern coal-fired electric plants remove much of the waste through filtration, but these wastes have to be dealt with so they don't leach heavy metals into the ground or into the drinking water supply.

The mining of coal also poses numerous problems and West Virginia serves as prime example. In order to extract the coal, which is often buried deep underground, large areas of mountains are removed, or large pits are dug. The major problem with mining occurs when tailings from these large-scale operations are dumped into streams polluting them with heavy metals. Environmental degradation is not limited to coal mining.

Acid mine drainage from mines that target gold, silver, copper, or other metals, is also a severe problem; especially throughout the west where the lower pH kills fish and invertebrates.

Untreated sewage creates another major problem, especially in poor regions around the world. It is well known that drinking contaminated water leads to several health problems in people. But, it also causes problems for the rivers, lakes, and marine environments. Earlier, we learned that aquatic ecosystems are limited by nutrients. Untreated sewage entering our water ways serves as a major source of nutrients directly leading to algae blooms. In aquatic systems, large algae blooms quickly lead to eutrophic conditions causing large fish kills. Fertilizer runoff from farms also causes algal blooms and dead zones. A striking example of a dead zone occurs in the Gulf of Mexico where the Mississippi River brings tons of nutrients to region.

To the Future

No matter what we do, humans will impact the world. The most pressing problem facing our civilization and the Earth's diversity comes from rapid population growth, causing the problems I previously discussed. Unfortunately, our economic models are based on continuous population growth. Every month a job reports comes out reporting the addition (or loss) of new jobs. A growing economy is great because it generates wealth and improves our standard of living.

Unfortunately, the population cannot keep growing indefinitely, there are just not enough resources on this planet to sustain a continually growing population. Confounding the problem of rapid growth lies in people's attitude and beliefs where most people are vehemently against birth restrictions. However, most economists agree that conservation is better for the economy in terms of job growth and higher earnings in the long run. A commercial fisherman may lose his job as a fisherman, but he may make more money in ecotourism by re-purposing his boat from fishing to accommodating divers, whale-watchers, or birders. A coal miner may earn the same amount of money manufacturing solar panels, but will be healthier from not having to work in harsh conditions.

As the human population continues to grow and the world becomes further degraded, biodiversity will continue to decline. We are simplifying the Earth's ecosystems through urbanization, clear cutting, habitat fragmentation, disrupting natural disturbance regimes, and spreading invasive species. All of these factors contribute to the world-wide decline in biodiversity as it becomes increasingly homogenized. As ecosystems become increasingly fragmented and degraded, fewer native species will remain. Eventually, specialized species dependent on unique habitats will be replaced by a few generalist species capable of living with humans or living in disturbed habitats.

Climate change is worsening the decline in biodiversity caused by habitat loss and the spread of invasive species. Native species must exist within their fundamental niche; as the climate changes they will have to move to find suitable habitat, evolve to keep pace with climate change, or go extinct. Unfortunately, many species can't move to suitable habitat if the climate changes because a once continuous habitat has been chopped into many smaller fragments. And, current nature preserves may have their ecosystems altered as a result of climate change and the spread of invasive species, making them unsuitable for the species they were meant to preserve.

Importantly, climate change will also affect our civilization. As a species we are young, about 200,000 years old, civilization is less than 10,000 years old. The rise of civilization was brought about by the agricultural revolution, coinciding with a period of very stable climate. By having a stable climate, our ancestors could predict when to plant and harvest their crops. Global climate change threatens the stability of our climate, which could make it hard to grow the crops we need to feed everyone. A 20-foot rise in sea levels would displace hundreds of millions of people from coastal regions. States like Florida will become flooded causing major economic losses for the region. Intensification of weather patterns will bring longer prolonged drought as we have seen in California, or an increase in severe thunderstorms, or hurricanes.

When it comes to climate change, less industrialized nations wish to improve their economies by utilizing coal and oil because they are cheap sources of energy. However, the long-term health of their people will suffer by creating more pollution, also leading to further loss of diversity and ecosystem functioning. Although, switching to renewable resources would be more beneficial in the long run as less pollution would be generated leading to fewer environmental problems.

We are clearly causing the 6th mass extinction. Time will tell the extent of the loss of diversity. But, life will go on, humans will most likely continue to exists for some time on this planet; although most likely not with 7-8 billion people. It will take millions of years for biodiversity to recover to its previous levels before the arrival of humans. Perhaps we are entering a new era, we could call it the Homogenozoic Era. Here's why I believe we are entering a new geological era, and why we should call it the Homogenozoic.

So far, the Phanerozoic eon has been divided into three eras, the first two ended with a mass extinction that paved the way for new lineages to become dominant. The end of the Paleozoic Era ended with a mass extinction that wiped out 90% of all life, including the dominant animals on land paving the way for the dinosaurs and other reptiles to rule the Mesozoic. The Mesozoic Era ended with the extinction of the dinosaurs and nearly 80% of all species paving the way for mammals and birds.

At our current rate, we will match those losses. But, this time the extinction is different, it was caused by the presence of another species, *Homo sapiens*. The name Homogenozoic comes from the fact that we are not only causing a mass extinction, we have moved species all over the planet, which will have a lasting effect on the future diversity of the planet.

Prior to humans moving animals around the planet, animals and plants had certain biogeographical patterns due to the history of plate tectonics. For example, Australia, known for its marsupials, broke off from Pangaea and other continents before eutherians, or modern mammals with a more developed placenta and longer gestation evolved. New Zealand is a continental fragment that prior to humans didn't have mammals, except for bats which flew there. Each biogeographical region had its own unique lineages. However, we have moved thousands of species around the planet, and many to areas they could never have reached due to natural boundaries like oceans and mountain ranges.

Many of these invasive species thrive in the disturbed habitats created by us. House Sparrows and Pigeons are found in most cities worldwide. European Starlings were introduced the US in the late 1800s and have spread throughout the continent becoming one of the most abundant and widespread birds in the continent. Salt Cedar became dominant in wetland areas throughout the southwest. Meanwhile, most other species are rapidly declining. As they wink out of existence, it will be the invasive species that survived humans which will be left to repopulate the Earth and evolve new species.

If you were to visit the planet 100 million years from now, diversity would have recovered, but the current mass extinction would stand out in the fossil record. But, the survivors, today's invasive species would not follow historical patterns of biogeography. To these future scientists, or aliens, it would appear that the world's diversity was homogenized. Based on this, I say the Homogenozoic began with the spread of people across the Pacific beginning the first wave of migration bringing rats, cats, dogs, and farm animals to remote islands causing the first round of extinction by the introduction of invasive species.

Beginning the Homogenozoic Era, I'd name the first period the Anthropocene, named after the effects of man that caused one of the largest changes to life on the Earth in more than a quarter of a billion years, or perhaps even more! Who knows what future scientists will say about the time period we live in now. Hopefully, as a society we will make wise choices to prevent the worst-case scenarios of climate change and losses of diversity.

Life has come a long way since its ancient origins. It took over 3.8 billion years for the first truly sentient species to evolve. Our advanced technology has made life easier for us, greatly expanded our knowledge of the universe, but has come at a price as harm our planet. Not to sound bleak, but much of the damage to our planet has already been done, and set us on a course of rapid climate change and collapsing ecosystems leading to a great mass extinction.

We can attempt to mitigate the severity of the problems by recycling, consuming less, conserving our resources, and voting for politicians that wish to serve our interest. But, one thing we have learned about life, it has a tenacious ability to adapt and evolve. Ironically, the same cosmopolitan invasive species that we have spread around the Earth, will likely be the ancestors of future species. Our activities will leave a permanent mark on the Earth for millions of years. We have permanently altered the course of evolution for all life on this planet.

What about the fate of humanity? Well, here is where I really become speculative, so feel free to disagree with me. Maybe, I'll at least spark your imagination. My hypothesis is that humans, or at least our descendants, will exists on this planet for a very long time, tens of millions of years, perhaps until the sun ends its life more than 5 billion years in the future. Although, by then our future descendants may not be quite human as they will have been evolving for a long time.

You may be wondering why I used the term hypothesis rather than opinion. Basically, it meets the requirements of a scientific hypothesis, it is testable and potentially falsifiable, although I will likely never know the answer. I base my hypothesis on observations of other long-lived species. First, we are widespread and found in a variety of habitats. Second, we are not dependent on a single resource or a limited number of resources for survival. Additionally, and perhaps most importantly, we no longer have to rely on evolution for us to adapt to a changing environment. Our unique intelligence allows us to quickly acclimate to a given situation. No other species has ever had this unique advantage. However, the continuation of our advanced civilization and high standard of living is totally dependent on cheap energy and creating a balance between population size and sustainable use of our resources.

www.ingramcontent.com/pod-product-compliance
Lightning Source LLC
Chambersburg PA
CBHW071013280326
41935CB00011B/1339